An Introduction to Proof via Inquiry-Based Learning

AMS/MAA | SPECTRUM

VOL 73

An Introduction to Proof via Inquiry-Based Learning

Dana C. Ernst

Providence, Rhode Island

2020 *Mathematics Subject Classification*. Primary 00-01, 03-01.

For additional information and updates on this book, visit
www.ams.org/bookpages/text-73

Library of Congress Cataloging-in-Publication Data

Names: Ernst, Dana C., 1975– author.
Title: An introduction to proof via inquiry-based learning / Dana C. Ernst.
Description: Providence, Rhode Island : MAA Press, an imprint of the American Mathematical Society, [2022] | Series: AMS/MAA textbooks, 2577-1205 ; volume 73
Identifiers: LCCN 2022002048 | ISBN 9781470463335 (paperback)
Subjects: LCSH: Proof theory–Textbooks. | Inquiry-based learning. | AMS: General – Instructional exposition (textbooks, tutorial papers, etc.) pertaining to mathematics in general. | Mathematical logic and foundations – Instructional exposition (textbooks, tutorial papers, etc.) pertaining to mathematical logic and foundations.
Classification: LCC QA9.54 .E76 2022 | DDC 511.3/6–dc23/eng20220318
LC record available at https://lccn.loc.gov/2022002048

10 9 8 7 6 5 4 3 2 1 27 26 25 24 23 22

Contents

Preface

Mathematics is not about calculations, but ideas. My goal as a teacher is to provide students with the opportunity to grapple with these ideas and to be immersed in the process of mathematical discovery. Repeatedly engaging in this process hones the mind and develops mental maturity marked by clear and rigorous thinking. Like music and art, mathematics provides an opportunity for enrichment, experiencing beauty, elegance, and aesthetic value. The medium of a painter is color and shape, whereas the medium of a mathematician is abstract thought. The creative aspect of mathematics is what captivates me and fuels my motivation to keep learning and exploring.

While the content we teach our students is important, it is not enough. An education must prepare individuals to ask and explore questions in contexts that do not yet exist and to be able to tackle problems they have never encountered. It is important that we put these issues front and center and place an explicit focus on students producing, rather than consuming, knowledge. If we truly want our students to be independent, inquisitive, and persistent, then we need to provide them with the means to acquire these skills. Their viability as professionals in the modern workforce depends on their ability to embrace this mindset.

When I started teaching, I mimicked the experiences I had as a student. Because it was all I knew, I lectured. By standard metrics, this seemed to work out just fine. Glowing student and peer evaluations, as well as reoccurring teaching awards, indicated that I was effectively doing my job. People consistently told me that I was an excellent teacher. However, two observations made me reconsider how well I was really doing. Namely, many of my students seemed to depend on me to be successful, and second, they retained only some of what I had taught them. In the words of Dylan Retsek:

> Things my students claim that I taught them masterfully, they don't know.

Inspired by a desire to address these concerns, I began transitioning away from direct instruction towards a more student-centered approach. The goals and philosophy behind inquiry-based learning (IBL) resonate deeply with my ideals, which is why I have embraced this paradigm. According to the Academy of Inquiry-Based Learning, IBL is a method of teaching that engages students in sense-making activities. Students are given tasks requiring them to solve problems, conjecture, experiment, explore, create, and communicate—all those wonderful skills and habits of mind that mathematicians engage in regularly. This book has IBL baked into its core.

This book is intended to be a task sequence for an introduction to proof course that utilizes an IBL approach. The primary objectives of this book are to:

- Expand the mathematical content knowledge of the reader,
- Provide an opportunity for the reader to experience the profound beauty of mathematics,
- Allow the reader to exercise creativity in producing and discovering mathematics,
- Enhance the ability of the reader to be a robust and persistent problem solver.

Ultimately, this is really a book about productive struggle and learning how to learn. Mathematics is simply the vehicle.

You can find the most up-to-date version of this textbook on GitHub:

`http://dcernst.github.io/IBL-IntroToProof/`

I would be thrilled if you used this textbook and improved it. If you make any modifications, you can either make a pull request on GitHub or submit the improvements via email. You are also welcome to fork the source and modify the text for your purposes as long as you maintain the Creative Commons Attribution–Share Alike 4.0 International License.

> Much more important than specific mathematical results are the habits of mind used by the people who create those results. ...Although it is necessary to infuse courses and curricula with modern content, what is even more important is to give students the tools they will need in order to use, understand, and even make mathematics that does not yet exist.
>
> Cuoco, Goldenberg, & Mark in *Habit of Mind: An Organizing Principle for Mathematics Curriculum*

Acknowledgments

The first draft of this book was written in 2009. At that time, several of the sections were adaptations of course materials written by Matthew Jones (CSU Dominguez Hills) and Stan Yoshinobu (University of Toronto). The current version of the book is the result of many iterations that involved the addition of new material, retooling of existing sections, and feedback from instructors who have used the book. The current version of the book is a far cry from what it looked like in 2009.

This book has been an open-source project since day one. Instructors and students can download the PDF for free and modify the source as they see fit. Several instructors and students have provided extremely useful feedback, which has improved the book at each iteration. Moreover, due to the open-source nature of the book, I have been able to incorporate content written by others. Below is a partial list of people (alphabetical by last name) who have contributed content, advice, or feedback.

- Chris Drupieski, T. Kyle Petersen, and Bridget Tenner (DePaul University). Modifications that these three made to the book inspired me to streamline some of the exposition, especially in the early chapters.

- Paul Ellis (Manhattanville College). Paul has provided lots of useful feedback and several suggestions for improvements. Paul suggested problems for Chapter 4 and provided an initial draft of Section 8.4: Images and Preimages of Functions.

- Jason Grout (Bloomberg, L.P.). I am extremely grateful to Jason for feedback on early versions of this manuscript, as well as for helping me with a variety of technical aspects of writing an open-source textbook.

- Anders Hendrickson (Milliman). Anders is the original author of the content in Appendix A: Elements of Style for Proofs. The current version in Appendix A is a result of modifications made by myself with some suggestions from David Richeson.

- Rebecca Jayne (Hampden–Sydney College). The current version of Section 4.3: Complete Induction is a derivative of content originally contributed by Rebecca.
- Matthew Jones (CSU Dominguez Hills) and Stan Yoshinobu (University of Toronto). A few of the sections were originally adaptations of notes written by Matt and Stan. Early versions of this textbook relied heavily on their work. Moreover, Matt and Stan were two of the key players that contributed to shaping my approach to teaching.
- David Richeson (Dickinson College). David is responsible for much of the content in Appendix B: Fancy Mathematical Terms, Appendix C: Paradoxes, and Appendix D: Definitions in Mathematics. In addition, the current version of Chapter 6: Three Famous Theorems is heavily based on content contributed by David.
- Carol Schumacher (Kenyon College). When I was transitioning to an IBL approach to teaching, Carol was one of my mentors and played a significant role in my development as a teacher. Moreover, this work is undoubtably influenced my Carol's excellent book *Chapter Zero: Fundamental Notions of Advanced Mathematics*, which I used when teaching my very first IBL course.
- Josh Wiscons (CSU Sacramento). The current version of Section 7.4: Modular Arithmetic is a derivative of content contributed by Josh.

1

> The mathematician does not study pure mathematics because it is useful; he studies it because he delights in it, and he delights in it because it is beautiful.
>
> Henri Poincaré, mathematician & physicist

Introduction

1.1 What Is this Book All About?

This book is intended to be used for a one-semester/quarter introduction to proof course (sometimes referred to as a transition to proof course). The purpose of this book is to introduce the reader to the process of constructing and writing formal and rigorous mathematical proofs. The intended audience is mathematics majors and minors. However, this book is also appropriate for anyone curious about mathematics and writing proofs. Most users of this book will have taken at least one semester of calculus, although other than some familiarity with a few standard functions in Chapter 8, content knowledge of calculus is not required. The book includes more content than one can expect to cover in a single semester/quarter. This allows the instructor/reader to pick and choose the sections that suit their needs and desires. Each chapter takes a focused approach to the included topics, but also includes many gentle exercises aimed at developing intuition.

The following sections form the core of the book and are likely the sections that an instructor would focus on in a one-semester introduction to proof course.

- Chapter 2: Mathematics and Logic. All sections.
- Chapter 3: Set Theory. Sections 3.1, 3.3, 3.4, and 3.5.
- Chapter 4: Induction. All sections.
- Chapter 7: Relations and Partitions. Sections 7.1, 7.2, and 7.3.
- Chapter 8: Functions. Sections 8.1, 8.2, 8.3, and 8.4.
- Chapter 9: Cardinality. All sections.

Time permitting, instructors can pick and choose topics from the remaining sections. I typically cover the core sections listed above together with Chapter 6: Three Famous Theorems during a single semester. The *Instructor Guide* contains examples of a few possible paths through the material, as well as information about which sections and theorems depend on material earlier in the book.

> Mathematics, rightly viewed, possesses not only truth, but supreme beauty—a beauty cold and austere, like that of sculpture, without appeal to any part of our weaker nature, without the gorgeous trappings of painting or music, yet sublimely pure, and capable of a stern perfection such as only the greatest art can show. The true spirit of delight, the exaltation, the sense of being more than Man, which is the touchstone of the highest excellence, is to be found in mathematics as surely as poetry.
>
> Bertrand Russell, philosopher & mathematician

1.2 What Should You Expect?

Up to this point, it is possible that your experience of mathematics has been about using formulas and algorithms. You are used to being asked to do things like: "solve for x", "take the derivative of this function", "integrate this function", etc. Accomplishing tasks like these usually amounts to mimicking examples that you have seen in class or in your textbook. However, this is only one part of mathematics. Mathematicians experiment, make conjectures, write definitions, and prove theorems. While engaging with the material contained in this book, we will learn about doing all of these things, especially writing proofs. Mathematicians are in the business of proving theorems and this is exactly our endeavor. Ultimately, the focus of this book is on producing and discovering mathematics.

Your progress will be fueled by your ability to wrestle with mathematical ideas and to prove theorems. As you work through the book, you will find that you have ideas for proofs, but you are unsure of them. Do not be afraid to tinker and make mistakes. You can always revisit your work as you become more proficient. Do not expect to do most things perfectly on your first—or even second or third—attempt. The material is too rich for a human being to completely understand immediately. Learning a new skill requires dedication and patience during periods of frustration. Moreover, solving genuine problems is difficult and takes time. But it is also rewarding!

> You may encounter many defeats, but you must not be defeated.
>
> Maya Angelou, poet & activist

1.3 An Inquiry-Based Approach

In many mathematics classrooms, "doing mathematics" means following the rules dictated by the teacher, and "knowing mathematics" means remembering and applying them. However, this is not a typical mathematics textbook and is likely a significant departure from your prior experience, where mimicking prefabricated examples led you to success. In order to promote a more active participation in your learning, this book adheres to an educational philosophy called inquiry-based learning (IBL). IBL is a student-centered method of teaching that engages students in sense-making activities and challenges them to create or discover mathematics. In this book, you will be expected to actively engage with the topics at hand and to construct your own understanding. You will be given tasks requiring you to solve problems, conjecture, experiment, explore, create, and communicate. Rather than showing facts or a clear, smooth path to a solution, this book will guide and mentor you through an adventure in mathematical discovery.

This book makes no assumptions about the specifics of how your instructor chooses to implement an IBL approach. Generally speaking, students are told which problems and theorems to grapple with for the next class sessions, and then the majority of class time is devoted to students working in groups on unresolved solutions/proofs or having students present their proposed solutions/proofs to the rest of the class. Students should—as much as possible—be responsible for guiding the acquisition of knowledge and validating the ideas presented. That is, you should not be looking to the instructor as the sole authority. In an IBL course, instructor and students have joint responsibility for the depth and progress of the course. While effective IBL courses come in a variety of forms, they all possess a few essential ingredients. According to Laursen and Rasmussen (2019), the Four Pillars of IBL are:

- Students engage deeply with coherent and meaningful mathematical tasks.
- Students collaboratively process mathematical ideas.
- Instructors inquire into student thinking.
- Instructors foster equity in their design and facilitation choices.

This book can only address the first pillar while it is the responsibility of your instructor and class to develop a culture that provides an adequate environment for the remaining pillars to take root. If you are studying this material independent of a classroom setting, I encourage you to find a community where you can collaborate and discuss your ideas.

Just like learning to play an instrument or sport, you will have to learn new skills and ideas. Along this journey, you should expect a cycle of victory and defeat, experiencing a full range of emotions. Sometimes you will feel exhilarated, other times you might be seemingly paralyzed by extreme confusion. You will experience struggle and failure before you experience understanding. This is part of the normal learning process. If you are doing things well, you should be confused on a regular basis. Productive struggle and mistakes provide opportunities for growth. As the author of this text, I am here to guide and challenge you, but I cannot do the learning for you, just as a music teacher cannot move your fingers and your heart for you. This is a very exciting time in your mathematical career. You will experience mathematics in a new and profound way. Be patient with yourself and others as you adjust to a new paradigm.

You could view this book as mountaineering guidebook. I have provided a list of mountains to summit, sometimes indicating which trailhead to start at or which trail to follow. There will always be multiple routes to top, some more challenging than others. Some summits you will attain quickly and easily, others might require a multi-day expedition. Oftentimes, your journey will be laced with false summits. Some summits will be obscured by clouds. Sometimes you will have to wait out a storm, perhaps turning around and attempting another route, or even attempting to summit on a different day after the weather has cleared. The strength, fitness, and endurance you gain along the way will allow you to take on more and more challenging, and often beautiful, terrain. Do not forget to take in the view from the top! The joy you feel from overcoming obstacles and reaching each summit under your own will and power has the potential to be life changing. But make no mistake, the journey is vastly more important than the destinations.

> Don't fear failure. Not failure, but low aim, is the crime. In great attempts it is glorious even to fail.
>
> Bruce Lee, martial artist & actor

1.4 Structure of the Textbook

As you read this book, you will be required to digest the material in a meaningful way. It is your responsibility to read and understand new definitions and their related concepts. In addition, you will be asked to complete problems aimed at solidifying your understanding of the material. Most importantly, you will be asked to make conjectures, produce counterexamples, and prove theorems. All of these tasks will almost always be challenging.

The items labeled as **Definition** and **Example** are meant to be read and digested. However, the items labeled as **Problem**, **Theorem**, and **Corollary** require action on your part. Items labeled as **Problem** are sort of a mixed bag. Some Problems are computational in nature and aimed at improving your understanding of a particular concept while others ask you to provide a counterexample for a statement if it is false or to provide a proof if the statement is true. Items with the **Theorem** and **Corollary** designation are mathematical facts and the intention is for you to produce a valid proof of the given statement. The main difference between a theorem and a corollary is that corollaries are typically statements that follow quickly from a previous theorem. In general, you should expect corollaries to have very short proofs. However, that does not mean that you cannot produce a more lengthy yet valid proof of a corollary.

Oftentimes, the problems and theorems are guiding you towards a substantial, more general result. Other times, they are designed to get you to apply ideas in a new way. One thing to always keep in mind is that every task in this book can be done by you, the student. But it may not be on your first try, or even your second.

Discussion of new topics is typically kept at a minimum and there are very few examples in this book. This is intentional. One of the objectives of the items labeled as **Problem** is for you to produce the examples needed to internalize unfamiliar concepts. The overarching goal of this book is to help you develop a deep and meaningful understanding of the processes of producing mathematics by putting you in direct contact with mathematical phenomena.

> Don't just read it; fight it! Ask your own questions, look for your own examples, discover your own proofs. Is the hypothesis necessary? Is the converse true? What happens in the classical special case? What about the degenerate cases? Where does the proof use the hypothesis?
>
> Paul Halmos, mathematician

1.5 Some Minimal Guidance

Especially in the opening sections, it will not be clear what facts from your prior experience in mathematics you are "allowed" to use. Unfortunately, addressing this issue is difficult and is something we will sort out along the way. In addition, you are likely unfamiliar with how to structure a valid mathematical proof. So that you do not feel completely abandoned, here are some guidelines to keep in mind as you get started with writing proofs.

- The statement you are proving should be on the same page as the beginning of your proof.
- You should indicate where the proof begins by writing "*Proof.*" at the beginning.
- Make it clear to yourself and the reader what your assumptions are at the very beginning of your proof. Typically, these statements will start off "Assume...", "Suppose...", or "Let...". Sometimes there will be some implicit assumptions that we can omit, but at least in the beginning, you should get in the habit of clearly stating your assumptions up front.
- Carefully consider the order in which you write your proof. Each sentence should follow from an earlier sentence in your proof or possibly a result you have already proved.
- Unlike the experience many of you had writing proofs in your high school geometry class, our proofs should be written in complete sentences. You should break sections of a proof into paragraphs and use proper grammar. There are some pedantic conventions for doing this that will be pointed out along the way. Initially, this will be an issue that you may struggle with, but you will get the hang of it.
- There will be many situations where you will want to refer to an earlier definition, problem, theorem, or corollary. In this case, you should reference the statement by number, but it is also helpful to the reader to summarize the statement you are citing. For example, you might write something like, "According to Theorem 2.3, the sum of two consecutive integers is odd, and so..." or "By the definition of divides (Definition 2.5), it follows that...". One thing worth pointing out is that if we are citing a definition, theorem, or problem by number, we should capitalize "Definition", "Theorem", or "Problem", respectively (e.g., "According to Theorem 2.3..."). Otherwise, we do not capitalize these words (e.g., "By the definition of divides...").
- There will be times when we will need to do some basic algebraic manipulations. You should feel free to do this whenever the need arises. But you should show sufficient work along the way. In addition, you should organize your calculations so that each step follows from the previous. The order in which we write things matters. You do not need to write down justifications for basic algebraic manipulations (e.g., adding 1 to both sides of an equation, adding and subtracting the same amount on the same side of an equation, adding like terms, factoring, basic simplification, etc.).

- On the other hand, you do need to make explicit justification of the logical steps in a proof. As stated above, you should cite a previous definition, theorem, etc. when necessary.

- Similar to making it clear where your proof begins, you should indicate where it ends. It is common to conclude a proof with the standard "proof box" (□ or ■). This little square at end of a proof is sometimes called a **tombstone** or **Halmos symbol** after Hungarian-born American mathematician Paul Halmos (1916–2006).

It is of utmost importance that you work to understand every proof. Questions—asked to your instructor, your peers, and yourself—are often your best tool for determining whether you understand a proof. Another way to help you process and understand a proof is to try and make observations and connections between different ideas, proof statements and methods, and to compare various approaches.

If you would like additional guidance before you dig in, look over the guidelines in Appendix A: Elements of Style for Proofs. It is suggested that you review this appendix occasionally as you progress through the book as some guidelines may not initially make sense or seem relevant. Be prepared to put in a lot of time and do all the work. Your effort will pay off in intellectual development. Now, go have fun and start exploring mathematics!

> Our greatest glory is not in never falling, but in rising every time we fall.
>
> ---
>
> Confucius, philosopher

> Pure mathematics is the poetry of logical ideas.
>
> Albert Einstein, theoretical physicist

2 Mathematics and Logic

Before you get started, make sure you have read Chapter 1, which sets the tone for the work we will begin doing here. In addition, you might find it useful to read Appendix A: Elements of Style for Proofs. As stated at the end of Section 1.5, you are encouraged to review this appendix occasionally as you progress through the book as some guidelines may not initially make sense or seem relevant.

2.1 A Taste of Number Theory

It is important to point out that we are diving in head first here. As we get started, we are going to rely on your intuition and previous experience with proofs. This is intentional. What you will likely encounter is a general sense of what a proof entails, but you may not be able to articulate the finer details that you do and do not comprehend. There are going to be some subtle issues that you will be confronted with and one of our goals will be to elucidate as many of them as possible. We need to calibrate and develop an intellectual need for structure. You are encouraged to just try your hand at writing proofs for the problems in this section without too much concern for whether you are "doing it the right way." In Section 2.2, we will start over and begin to develop a formal foundation for the material in the remainder of the book. Once you have gained more experience and a better understanding of what a proof entails, you should consider returning to this section and reviewing your first attempts at writing proofs. In the meantime, see what you can do!

In this section, we will introduce the basics of a branch of mathematics called **number theory**, which is devoted to studying the properties of the integers. The integers is the set of numbers given by

$$\mathbb{Z} := \{\ldots, -3, -2, -1, 0, 1, 2, 3, \ldots\}.$$

The collection of positive integers also have a special name. The set of **natural numbers** is given by

$$\mathbb{N} := \{1, 2, 3, \ldots\}.$$

Some mathematicians (set theorists, in particular) include 0 in $\mathbb{N}$, but this will not be our convention. If you look closely at the two sets we defined above, you will notice that we wrote $:=$ instead of $=$. We use $:=$ to mean that the symbol or expression on the left is defined to be equal to the expression on the right. The symbol $\mathbb{R}$ is used to denote the set of all **real numbers**. We will not formally define the real numbers, but instead rely on your prior intuition and understanding.

Because you are so familiar with many of the properties of the integers and real numbers, one of the issues that we will bump into is knowing which facts we can take for granted. As a general rule of thumb, you should attempt to use the definitions provided without relying too much on your prior knowledge. The order in which we develop things is important.

It is common practice in mathematics to use the symbol $\in$ as an abbreviation for the phrase "is an element of" or sometimes simply "in." For example, the mathematical expression "$n \in \mathbb{Z}$" means "n is an element of the integers." However, some care should be taken in how this symbol is used. We will only use the symbol "$\in$" in expressions of the form $a \in A$, where A is a set and a is an element of A. We will write expressions like $a, b \in A$ as shorthand for "$a \in A$ and $b \in A$." We should avoid writing phrases such as "a is a number $\in A$" and "$n \in$ integers".

We will now encounter our very first definition. In mathematics, a **definition** is a precise and unambiguous description of the meaning of a mathematical term. It characterizes the meaning of a word by giving all the properties and only those properties that must be true. Check out Appendix B for a list of other mathematical terms that we should be familiar with.

Definition 2.1. An integer n is **even** if $n = 2k$ for some $k \in \mathbb{Z}$. An integer n is **odd** if $n = 2k + 1$ for some $k \in \mathbb{Z}$.

Notice that we framed the definition of "even" in terms of multiplication as opposed to division. When tackling theorems and problems involving even or odd, be sure to make use of our formal definitions and not some of the well-known divisibility properties. For now, you should avoid arguments that involve statements like, "even numbers have no remainder when divided by two" or "the last digit of an even number is 0, 2, 4, 6, or 8." Also notice that the notions of even and odd apply to zero and negative numbers. In particular, zero is even since $0 = 2 \cdot 0$, where it is worth emphasizing that the occurrence of 0 on the righthand side of the equation is an integer. As another example, we see that -1 is odd since $-1 = 2(-1) + 1$. Despite the fact that $-1 = 2(-1/2)$, this does not imply that -1 is also even since $-1/2$ is not an integer. For the remainder of this section, you may assume that every integer is either even or odd but never both.

Our first theorem concerning the integers is stated below. A **theorem** is a mathematical statement that is proved using rigorous mathematical reasoning. As with most theorems in this book, your task is to try your hand at proving the following theorem. Give it a try.

Theorem 2.2. *If n is an even integer, then n^2 is an even integer.*

One crux in proving the next theorem involves figuring out how to describe an arbitrary pair of consecutive integers.

Theorem 2.3. *The sum of two consecutive integers is odd.*

One skill we will want to develop is determining whether a given mathematical statement is true or false. In order to verify that a mathematical statement is false, we should provide a specific example where the statement fails. Such an example is called a **counterexample**. Notice that it is sufficient to provide a single example to verify that a general statement is not true. On the other hand, if we want to prove that a general mathematical statement is true, it is usually not sufficient to provide just a single example, or even a hundred examples. Such examples are just evidence that the statement is true.

Problem 2.4. Determine whether each of the following statements is true or false. If a statement is true, prove it. If a statement is false, provide a counterexample.

(a) The product of an odd integer and an even integer is odd.

(b) The product of an odd integer and an odd integer is odd.

(c) The product of an even integer and an even integer is even.

(d) The sum of an even integer and an odd integer is odd.

For the statements that were true in the previous problem, you may cite them later in a future proof as if they are theorems.

Definition 2.5. Given $n, m \in \mathbb{Z}$, we say that n **divides** m, written $\boxed{n|m}$, if there exists $k \in \mathbb{Z}$ such that $m = nk$. If $n|m$, we may also say that m is **divisible by** n or that n is a **factor** of m.

Problem 2.6. For $n, m \in \mathbb{Z}$, how are the following mathematical expressions similar and how are they different? In particular, is each one a sentence or simply a noun?

(a) $n|m$

(b) $\dfrac{m}{n}$

(c) m/n

In this section on number theory, we allow addition, subtraction, and multiplication of integers. In general, we avoid division since an integer divided by an integer may result in a number that is not an integer. The upshot is that we will avoid writing $\frac{m}{n}$. When you feel the urge to divide, switch to an equivalent formulation using multiplication. This will make your life much easier when proving statements involving divisibility.

Theorem 2.7. *The sum of any three consecutive integers is always divisible by three.*

Problem 2.8. Let $a, b, n, m \in \mathbb{Z}$. Determine whether each of the following statements is true or false. If a statement is true, prove it. If a statement is false, provide a counterexample.

(a) If $a|n$, then $a|mn$.

(b) If 6 divides n, then 2 divides n and 3 divides n.

(c) If ab divides n, then a divides n and b divides n.

A theorem that follows almost immediately from another theorem is called a **corollary**. See if you can prove the next result quickly using a previous result. Be sure to cite the result in your proof.

Corollary 2.9. *If $a, n \in \mathbb{Z}$ such that a divides n, then a divides n^2.*

The next two theorems are likely familiar to you.

Theorem 2.10. *If $a, n \in \mathbb{Z}$ such that a divides n, then a divides $-n$.*

Theorem 2.11. *If $a, n, m \in \mathbb{Z}$ such that a divides m and a divides n, then a divides $m + n$.*

Notice that we have been tinkering with statements of the form “If…, then…”. Statements of this form are called **conditional propositions**, which we revisit in the next section. The phrase that occurs after “If” but before “then” is called the **hypothesis** while the phrase that occurs after "then" is called the **conclusion**. For example, in Problem 2.8(a), "$a|n$" is the hypothesis while "$a|mn$" is the conclusion. Note that conditional propositions can also be written in the form "...if...", where the conclusion is written before "if" and the hypothesis after. For example, we can rewrite Problem 2.8(a) as "$a|mn$ if $a|n$". While the order of the hypothesis and conclusion have been reversed in the sentence, their roles have not.

Whenever we encounter a conditional statement in mathematics, we want to get in the habit of asking ourselves what happens when we swap the roles of the hypothesis and the conclusion. The statement that results from reversing the roles of the hypothesis and conclusion in a conditional statement is called the **converse** of the original statement. For example, the converse of Problem 2.8(a) is "If $a|mn$, then $a|n$", which happens to be false. The converse of Theorem 2.2 is "If n^2 is an even integer, then n is an even integer". While this statement is true it does *not* have the same meaning as Theorem 2.2.

Problem 2.12. Determine whether the converse of each of Corollary 2.9, Theorem 2.10, and Theorem 2.11 is true. That is, for $a, n, m \in \mathbb{Z}$, determine whether each of the following statements is true or false. If a statement is true, prove it. If a statement is false, provide a counterexample.

(a) If a divides n^2, then a divides n. (Converse of Corollary 2.9)

(b) If a divides $-n$, then a divides n. (Converse of Theorem 2.10)

(c) If a divides $m + n$, then a divides m and a divides n. (Converse of Theorem 2.11)

The next theorem is often referred to as the **transitivity of division of integers**.

Theorem 2.13. *If $a, b, c \in \mathbb{Z}$ such that a divides b and b divides c, then a divides c.*

Once we have proved a few theorems, we should be on the look out to see if we can utilize any of our current results to prove new results. There is no point in reinventing the wheel if we do not have to.

Theorem 2.14. *If $a, n, m \in \mathbb{Z}$ such that a divides m and a divides n, then a divides $m - n$.*

Theorem 2.15. *If $n \in \mathbb{Z}$ such that n is odd, then 8 divides $n^2 - 1$.*

> Time spent thinking about a problem is always time well spent. Even if you seem to make no progress at all.
>
> Paul Zeitz, mathematician

2.2 Introduction to Logic

In the previous section, we jumped in head first and attempted to prove several theorems in the context of number theory without a formal understanding of what it was we were doing. Likely, many issues bubbled to the surface. What is a proof? What sorts of statements require proof? What should a proof entail? How should a proof be structured? Let's take a step back and do a more careful examination of what it is we are actually doing. In the the next two sections, we will introduce the basics of **propositional logic**—also referred to as **propositional calculus** or sometimes **zeroth-order logic**.

Definition 2.16. A **proposition** is a sentence that is either true or false but never both. The **truth value** (or **logical value**) of a proposition refers to its attribute of being true or false.

For example, the sentence "All dogs have four legs" is a false proposition. However, the perfectly good sentence "$x = 1$" is *not* a proposition all by itself since we do not actually know what x is.

Problem 2.17. Determine whether each of the following is a proposition. Explain your reasoning.

(a) All cars are red.

(b) Every person whose name begins with J has the name Joe.

(c) $x^2 = 4$.

(d) There exists a real number x such that $x^2 = 4$.

(e) For all real numbers x, $x^2 = 4$.

(f) $\sqrt{2}$ is an irrational number.

(g) p is prime.

(h) Is it raining?

(i) It will rain tomorrow.

(j) Led Zeppelin is the best band of all time.

The last two sentences in the previous problem may stir debate. It is not so important that we come to consensus as to whether either of these two sentences is actually a proposition or not. The good news is that in mathematics we do not encounter statements whose truth value is dependent on either the future or opinion.

Given two propositions, we can form more complicated propositions using **logical connectives**.

Definition 2.18. Let A and B be propositions.

(a) The proposition "**not** A" is true if A is false; expressed symbolically as $\boxed{\neg A}$ and called the **negation** of A.

(b) The proposition "A **and** B" is true if both A and B are true; expressed symbolically as $\boxed{A \wedge B}$ and called the **conjunction** of A and B.

(c) The proposition "A **or** B" is true if at least one of A or B is true; expressed symbolically as $\boxed{A \vee B}$ and called the **disjunction** of A and B.

(d) The proposition "**If** A, **then** B" is true if both A and B are true, or A is false; expressed symbolically as $\boxed{A \implies B}$ and called a **conditional proposition** (or **implication**). In this case, A is called the **hypothesis** and B is called the **conclusion**. Note that $A \implies B$ may also be read as "A implies B", "A only if B", "B if A", or "B whenever A".

(e) The proposition "A **if and only if** B" (alternatively, "A **is necessary and sufficient for** B") is true if both A and B have the same truth value; expressed symbolically as $\boxed{A \iff B}$ and called a **biconditional proposition**. If $A \iff B$ is true, we say that A and B are **logically equivalent**.

Each of the boxed propositions is called a **compound proposition**, where A and B are referred to as the **components** of the compound proposition.

It is worth pointing out that definitions in mathematics are typically written in the form "B if A" (or "B provided that A" or "B whenever A"), where B contains the term or phrase we are defining and A provides the meaning of the concept we are defining. In the case of definitions, we should always interpret "B if A" as describing precisely the collection of "objects" (e.g., numbers, sets, functions, etc.) that should be identified with the term or phrase we defining. That is, if an object does not meet the condition specified in A, then it is never referred to by the term or phrase we are defining. Some authors will write definitions in the form "B if and only if A". However, a definition is not at all the same kind of statement as a usual biconditional since one of the two sides is undefined until the definition is made. A definition is really a statement that the newly defined term or phrase is synonymous with a previously defined concept.

We can form complicated compound propositions with several components by utilizing logical connectives.

Problem 2.19. Let A represent "6 is an even integer" and B represent "4 divides 6." Express each of the following compound propositions in an ordinary English sentence and then determine its truth value.

(a) $A \wedge B$

(b) $A \vee B$

(c) $\neg A$

(d) $\neg B$

(e) $\neg(A \wedge B)$

(f) $\neg(A \vee B)$

(g) $A \implies B$

Definition 2.20. A **truth table** for a compound proposition is a table that illustrates all possible combinations of truth values for the components of the compound proposition together with the resulting truth value for each combination.

Example 2.21. If A and B are propositions, then the truth table for the compound proposition $A \wedge B$ is given by the following.

A	B	$A \wedge B$
T	T	T
T	F	F
F	T	F
F	F	F

Notice that we have columns for each of A and B. The rows for these two columns correspond to all possible combinations of truth values for A and B. The third column yields the truth value of $A \wedge B$ given the possible truth values for A and B.

Each component of a compound proposition has two possible truth values, namely true or false. Thus, if a compound proposition is built from n component propositions, then the truth table will require 2^n rows.

Problem 2.22. Create a truth table for each of the following compound propositions. You should add additional columns to your tables as needed to assist you with intermediate steps. For example, you might need four columns for the third and fourth compound propositions below.

(a) $\neg A$

(b) $A \vee B$

(c) $\neg(A \wedge B)$

(d) $\neg A \wedge \neg B$

Problem 2.23. A coach promises her players, "If we win tonight, then I will buy you pizza tomorrow." Determine the cases in which the players can rightly claim to have been lied to. If the team lost the game and the coach decided to buy them pizza anyway, was she lying?

Problem 2.24. Use Definition 2.18(d) to construct a truth table for $A \implies B$. Compare your truth table with Problem 2.23. The combination you should pay particular attention to is when the hypothesis is false while the conclusion is true.

In accordance with Definition 2.18(d), a conditional proposition $A \implies B$ is only false when the hypothesis is true and the conclusion is false. Perhaps you are bothered by the fact that $A \implies B$ is true when A is false no matter what the truth value of B is. The thing to keep in mind is that the truth value of $A \implies B$ relies on a very specific definition and may not always agree with the colloquial use of "If..., then..." statements that we encounter in everyday language. For example, if someone says, "If you break the rules, then you will be punished", the speaker likely intends the statement to be interpreted as "You will be punished if and only if you break the rules." In logic and mathematics, we aim to remove such ambiguity by explicitly saying exactly what we mean. For our purposes, we should view a conditional proposition as a contract or obligation. If the hypothesis is false and the conclusion is true, the contract is not violated. On the other hand, if the hypothesis is true and the conclusion is false, then the contract is broken.

We can often prove facts concerning logical statements using truth tables. Recall that two propositions P and Q (both of which might be complicated compound propositions) are logically equivalent if $P \iff Q$ is true (see Definition 2.18(e)). This happens when P and Q have the same truth value. We can verify whether P and Q have the same truth value by constructing a truth table that includes columns for each of the components of P and Q, listing all possible combinations of their truth values, together with columns for P and Q that lists their resulting truth values. If the truth values in the columns for P and Q agree, then P and Q are logically equivalent, and otherwise they are not logically equivalent. When constructing truth tables to verify whether P and Q are logically equivalent, you should add any necessary intermediate columns to aid in your "calculations". Use truth tables when attempting to justify the next few problems.

Theorem 2.25. *If A is a proposition, then $\neg(\neg A)$ is logically equivalent to A.*

The next theorem, referred to as **De Morgan's Law**, provides a method for negating a compound proposition involving a conjunction.

Theorem 2.26 (De Morgan's Law)**.** *If A and B are propositions, then $\neg(A \wedge B)$ is logically equivalent to $\neg A \vee \neg B$.*

Problem 2.27 (De Morgan's Law)**.** Let A and B be propositions. Conjecture a statement similar to Theorem 2.26 for the proposition $\neg(A \vee B)$ and then prove it. This is also called De Morgan's Law.

We will make use of both versions De Morgan's Law on on a regular basis. Sometimes conjunctions and disjunctions are "buried" in a mathematical statement, which makes negating statements tricky business. Keep this in mind when approaching the next problem.

Problem 2.28. Let x be your favorite real number. Negate each of the following statements. Note that the statement in Part (b) involves a conjunction.

(a) $x < -1$ or $x \geq 3$.

(b) $0 \leq x < 1$.

Theorem 2.29. *If A and B are propositions, then $A \iff B$ is logically equivalent to $(A \implies B) \wedge (B \implies A)$.*

Theorem 2.30. *If A, B, and C are propositions, then $(A \vee B) \implies C$ is logically equivalent to $(A \implies C) \wedge (B \implies C)$.*

We already introduced the following notion in the discussion following Theorem 2.11

Definition 2.31. If A and B are propositions, then the **converse** of $A \implies B$ is $B \implies A$.

Problem 2.32. Provide an example of a true conditional proposition whose converse is false.

Definition 2.33. If A and B are propositions, then the **inverse** of $A \implies B$ is $\neg A \implies \neg B$.

Problem 2.34. Provide an example of a true conditional proposition whose inverse is false.

Based on Problems 2.32 and 2.34, we can conclude that the converse and inverse of a conditional proposition do not necessarily have the same truth value as the original statement. Moreover, the converse and inverse of a conditional proposition do not necessarily have the same truth value as each other.

Problem 2.35. If possible, provide an example of a conditional proposition whose converse is true but whose inverse is false. If this is not possible, explain why.

What if we swap the roles of the hypothesis and conclusion of a conditional proposition *and* negate each?

Definition 2.36. If A and B are propositions, then the **contrapositive** of $A \implies B$ is $\neg B \implies \neg A$.

Problem 2.37. Let A and B represent the statements from Problem 2.19. Express each of the following in an ordinary English sentence.

(a) The converse of $A \implies B$.

(b) The contrapositive of $A \implies B$.

Problem 2.38. Find the converse and the contrapositive of the following statement: "If Dana lives in Flagstaff, then Dana lives in Arizona."

Use a truth table to prove the following theorem.

Theorem 2.39. *If A and B are propositions, then $A \implies B$ is logically equivalent to its contrapositive.*

So far we have discussed how to negate propositions of the form A, $A \wedge B$, and $A \vee B$ for propositions A and B. However, we have yet to discuss how to negate propositions of the form $A \implies B$. Prove the following result with a truth table.

Theorem 2.40. *If A and B are propositions, then the implication $A \implies B$ is logically equivalent to the disjunction $\neg A \vee B$.*

The next result follows quickly from Theorem 2.40 together with De Morgan's Law. You can also verify this result using a truth table.

Corollary 2.41. *If A and B are propositions, then $\neg(A \implies B)$ is logically equivalent to $A \wedge \neg B$.*

Problem 2.42. Let A and B be the propositions "$\sqrt{2}$ is an irrational number" and "Every rectangle is a trapezoid," respectively.

(a) Express $A \implies B$ as an English sentence involving the disjunction "or."

(b) Express $\neg(A \implies B)$ as an English sentence involving the conjunction "and."

Problem 2.43. It turns out that the proposition "If $.\overline{99} = \frac{9}{10} + \frac{9}{100} + \frac{9}{1000} + \cdots$, then $.\overline{99} \neq 1$" is false. Write its negation as a conjunction.

Recall that a proposition is exclusively either true or false—it can never be both.

Definition 2.44. A compound proposition that is always false is called a **contradiction**. A compound proposition that is always true is called a **tautology**.

Theorem 2.45. *If A is a proposition, then the proposition $\neg A \wedge A$ is a contradiction.*

Problem 2.46. Provide an example of a tautology using arbitrary propositions and any of the logical connectives $\neg$, $\wedge$, and $\vee$. Prove that your example is in fact a tautology.

> I didn't want to just know names of things. I remember really wanting to know how it all worked.
>
> Elizabeth Blackburn, biologist

2.3 Techniques for Proving Conditional Propositions

Each of the theorems that we proved in Section 2.1 are examples of conditional propositions. However, some of the statements were disguised as such. For example, Theorem 2.3 states, "The sum of two consecutive integers is odd." We can reword this theorem as, "If $n \in \mathbb{Z}$, then $n + (n + 1)$ is odd."

Problem 2.47. Reword Theorem 2.7 so that it explicitly reads as a conditional proposition.

Each of the proofs that you produced in Section 2.1 had the same format, which we refer to as a **direct proof**.

Skeleton Proof 2.48 (Proof of $A \implies B$ by direct proof)**.** If you want to prove the implication $A \implies B$ via a direct proof, then the structure of the proof is as follows.

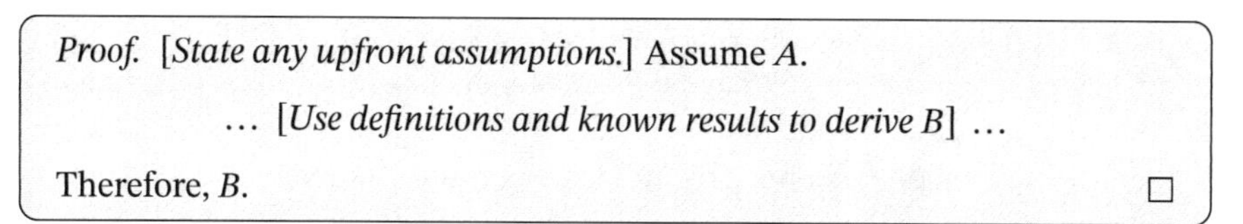

Proof. [*State any upfront assumptions.*] Assume A.

... [*Use definitions and known results to derive* B] ...

Therefore, B. □

Take a few minutes to review the proofs that you wrote in Section 2.1 and see if you can witness the structure of Skeleton Proof 2.48 in your proofs.

The upshot of Theorem 2.39 is that if you want to prove a conditional proposition, you can prove its contrapositive instead. This approach is called a **proof by contraposition**.

Skeleton Proof 2.49 (Proof of $A \implies B$ by contraposition). If you want to prove the implication $A \implies B$ by proving its contrapositive $\neg B \implies \neg A$ instead, then the structure of the proof is as follows.

Proof. [*State any upfront assumptions.*] We will utilize a proof by contraposition. Assume $\neg B$.

... [*Use definitions and known results to derive* $\neg A$] ...

Therefore, $\neg A$. We have proved the contrapositive, and hence if A, then B. □

We have introduced the logical symbols $\neg$, $\wedge$, $\vee$, $\implies$, and $\iff$ since it provides a convenient way of discussing the formality of logic. However, when writing mathematical proofs, you should avoid using these symbols.

Problem 2.50. Consider the following statement: If $x \in \mathbb{Z}$ such that x^2 is odd, then x is odd.

The items below can be assembled to form a proof of this statement, but they are currently out of order. Put them in the proper order.

(1) Assume that x is an even integer.

(2) We will utilize a proof by contraposition.

(3) Thus, x^2 is twice an integer.

(4) Since $x = 2k$, we have that $x^2 = (2k)^2 = 4k^2$.

(5) Since k is an integer, $2k^2$ is also an integer.

(6) By the definition of even, there is an integer k such that $x = 2k$.

(7) We have proved the contrapositive, and hence the desired statement is true.

(8) Assume $x \in \mathbb{Z}$.

(9) By the definition of even integer, x^2 is an even integer.

(10) Notice that $x^2 = 2(2k^2)$.

Prove the next two theorems by proving the contrapositive of the given statement.

Theorem 2.51. *If $n \in \mathbb{Z}$ such that n^2 is even, then n is even.*

Theorem 2.52. *If $n, m \in \mathbb{Z}$ such that nm is even, then n is even or m is even.*

Suppose that we want to prove some proposition P (which might be something like $A \implies B$ or even more complicated). One approach, called **proof by contradiction**, is to assume $\neg P$ and then logically deduce a contradiction of the form $Q \wedge \neg Q$, where Q is some proposition. Since this is absurd, the assumption $\neg P$ must have been false, so P is true. The tricky part about a proof by contradiction is that it is not usually obvious what the statement Q should be.

Skeleton Proof 2.53 (Proof of P by contradiction). Here is what the general structure for a proof by contradiction looks like if we are trying to prove the proposition P.

Proof. [*State any upfront assumptions.*] For sake of a contradiction, assume $\neg P$.

... [*Use definitions and known results to derive some Q and its negation $\neg Q$.*] ...

This is a contradiction. Therefore, P. □

Proof by contradiction can be useful for proving statements of the form $A \implies B$, where $\neg B$ is easier to "get your hands on," because $\neg(A \implies B)$ is logically equivalent to $A \wedge \neg B$ (see Corollary 2.41).

Skeleton Proof 2.54 (Proof of $A \implies B$ by contradiction). If you want to prove the implication $A \implies B$ via a proof by contradiction, then the structure of the proof is as follows.

Proof. [*State any upfront assumptions.*] For sake of a contradiction, assume A and $\neg B$.

... [*Use definitions and known results to derive some Q and its negation $\neg Q$.*] ...

This is a contradiction. Therefore, if A, then B. □

Problem 2.55. Assume that $x \in \mathbb{Z}$. Consider the following proposition: If x is odd, then 2 does not divide x.

(a) Prove the contrapositive of this statement.

(b) Prove the statement using a proof by contradiction.

Prove the following theorem via a proof by contradiction. Afterward, consider the difficulties one might encounter when trying to prove the result more directly. The given statement is not true if we replace $\mathbb{N}$ with $\mathbb{Z}$. Do you see why?

Theorem 2.56. *Assume that $x, y \in \mathbb{N}$. If x divides y, then $x \leq y$.*

Oftentimes a conditional proposition can be proved via a direct proof and by using a proof by contradiction. Most mathematicians view a direct proof to be more elegant than a proof by contradiction. When approaching the proof of a conditional proposition, you should strive for a direct proof. In general, if you are attempting to prove $A \implies B$ using a proof by contradiction and you end up with $\neg B$ and B (which yields a contradiction), then this is evidence that a proof by contradiction was unnecessary. On the other hand, if you end up with $\neg Q$ and Q, where Q is not the same as B, then a proof by contradiction is a reasonable approach.

In light of Theorem 2.29, if we want to prove a biconditional of the form $A \iff B$, we need to prove both $A \implies B$ and $B \implies A$. You should always make it clear to the reader when you are proving each implication. One approach is to label each subproof with "($\implies$)" and "($\impliedby$)" (including the parentheses), respectively. Occasionally, you will discover that the proof of one implication is exactly the reverse of the proof of the other implication. If this happens to be the case, you may skip writing two subproofs and simply write a single proof that chains together each step using biconditionals. Such proofs will almost always be shorter, but can be challenging to write in an eloquent way. It is always a safe bet to write a separate subproof for each implication.

When proving each implication of a biconditional, you may choose to utilize a direct proof, a proof by contraposition, or a proof by contradiction. For example, you could prove the first implication using a proof by contradiction and a direct proof for the second implication.

The following theorem provides an opportunity to gain some experience with writing proofs of biconditional statements.

Theorem 2.57. *Let $n \in \mathbb{Z}$. Then n is even if and only if 4 divides n^2.*

> Making learning easy does not necessarily ease learning.
>
> ---
>
> Manu Kapur, learning scientist

2.4 Introduction to Quantification

In this section and the next, we introduce **first-order logic**—also referred to as **predicate logic**, **quantificational logic**, and **first-order predicate calculus**. The sentence "$x > 0$" is not itself a proposition because its truth value depends on x. In this case, we say that x is a **free variable**. A sentence with at least one free variable is called a **predicate** (or **open sentence**). To turn a predicate into a proposition, we must either substitute values for each free variable or "quantify" the free variables. We will use notation such as $P(x)$ and $Q(a, b)$ to represent predicates with free variables x and a, b, respectively. The letters "P" and "Q" that we used in the previous sentence are not special; we can use any letter or symbol we want. For example, each of the following represents a predicate with the indicated free variables.

- $S(x) :=$ "$x^2 - 4 = 0$"
- $L(a, b) :=$ "$a < b$"
- $F(x, y) :=$ "x is friends with y"

Note that we used quotation marks above to remove some ambiguity. What would $S(x) = x^2 - 4 = 0$ mean? It looks like $S(x)$ equals 0, but actually we want $S(x)$ to represent the whole sentence "$x^2 - 4 = 0$". Also, notice that the order in which we utilize the free variables might matter. For example, compare $L(a, b)$ with $L(b, a)$.

One way we can make propositions out of predicates is by assigning specific values to the free variables. That is, if $P(x)$ is a predicate and x_0 is specific value for x, then $P(x_0)$ is now a proposition that is either true or false.

Problem 2.58. Consider $S(x)$ and $L(a, b)$ as defined above. Determine the truth values of $S(0)$, $S(-2)$, $L(2, 1)$, and $L(-3, -2)$. Is $L(2, b)$ a proposition or a predicate?

Besides substituting specific values for free variables in a predicate, we can also make a claim about which values of the free variables apply to the predicate.

Problem 2.59. Both of the following sentences are propositions. Decide whether each is true or false. What would it take to justify your answers?

(a) For all $x \in \mathbb{R}$, $x^2 - 4 = 0$.

(b) There exists $x \in \mathbb{R}$ such that $x^2 - 4 = 0$.

Definition 2.60. "For all" is the **universal quantifier** and "there exists...such that" is the **existential quantifier**.

In mathematics, the phrases "for all", "for any", "for every", and "for each" can be used interchangeably (even though they might convey slightly different meanings in colloquial language). We can replace "there exists...such that" with phrases like "for some" (possibly with some tweaking of the wording of the sentence). It is important to note that the existential quantifier is making a claim about "at least one", *not* "exactly one."

Variables that are quantified with a universal or existential quantifier are said to be **bound**. To be a proposition, *all* variables of a predicate must be bound.

We must take care to specify the collection of acceptable values for the free variables. Consider the sentence "For all x, $x > 0$." Is this sentence true or false? The answer depends on what set the universal quantifier applies to. Certainly, the sentence is false if we apply it for all $x \in \mathbb{Z}$. However, the sentence is true for all $x \in \mathbb{N}$. Context may resolve ambiguities, but otherwise, we must write clearly: "For all $x \in \mathbb{Z}, x > 0$" or "For all $x \in \mathbb{N}, x > 0$." The collection of intended values for a variable is called the **universe of discourse**.

Problem 2.61. Suppose our universe of discourse is the set of integers.

(a) Provide an example of a predicate $P(x)$ such that "For all x, $P(x)$" is true.

(b) Provide an example of a predicate $Q(x)$ such that "For all x, $Q(x)$" is false while "There exists x such that $Q(x)$" is true.

If a predicate has more than one free variable, then we can build propositions by quantifying each variable. However, *the order of the quantifiers is extremely important!*

Problem 2.62. Let $P(x, y)$ be a predicate with free variables x and y in a universe of discourse U. One way to quantify the variables is "For all $x \in U$, there exists $y \in U$ such that $P(x, y)$." How else can the variables be quantified?

The next problem illustrates that at least some of the possibilities you discovered in the previous problem are *not* equivalent to each other.

Problem 2.63. Suppose the universe of discourse is the set of people and consider the predicate $M(x, y) :=$" x is married to y". We can interpret the formal statement "For all x, there exists y such that $M(x, y)$" as meaning "Everybody is married to somebody." Interpret the meaning of each of the following statements in a similar way.

(a) For all x, there exists y such that $M(x, y)$.

(b) There exists y such that for all x, $M(x, y)$.

(c) For all x, for all y, $M(x, y)$.

(d) There exists x such that there exists y such that $M(x, y)$.

Problem 2.64. Suppose the universe of discourse is the set of real numbers and consider the predicate $F(x, y) :=" x = y^2"$. Interpret the meaning of each of the following statements.

(a) There exists x such that there exists y such that $F(x, y)$.

(b) There exists y such that there exists x such that $F(x, y)$.

(c) For all y, for all x, $F(x, y)$.

There are a couple of key points to keep in mind about quantification. To be a proposition, all variables must be quantified. This can happen in at least two ways:

- The variables are explicitly bound by quantifiers in the same sentence.
- The variables are implicitly bound by preceding sentences or by context. Statements of the form "Let $x = \ldots$" and "Assume $x \in \ldots$" bind the variable x and remove ambiguity.

Also, the order of the quantification is important. Reversing the order of the quantifiers can substantially change the meaning of a proposition.

Quantification and logical connectives ("and", "or", "If..., then...", and "not") enable complex mathematical statements. For example, if f is a function while c and L are real numbers, then the formal definition of $\lim_{x\to c} f(x) = L$, which you may have encountered in calculus, is:

> For all $\varepsilon > 0$, there exists $\delta > 0$ such that for all x, if $0 < |x - c| < \delta$, then $|f(x) - L| < \varepsilon$.

In order to study the abstract nature of complicated mathematical statements, it is useful to adopt some notation.

Definition 2.65. The universal quantifier "for all" is denoted $\forall$, and the existential quantifier "there exists...such that" is denoted $\exists$.

Using our abbreviations for the logical connectives and quantifiers, we can symbolically represent mathematical propositions. For example, the (true) proposition "There exists $x \in \mathbb{R}$ such that $x^2 - 1 = 0$" becomes "$(\exists x \in \mathbb{R})(x^2 - 1 = 0)$," while the (false) proposition "For all $x \in \mathbb{N}$, there exists $y \in \mathbb{N}$ such that $y < x$" becomes "$(\forall x \in \mathbb{N})(\exists y \in \mathbb{N})(y < x)$."

Problem 2.66. Convert the following propositions into statements using only logical and mathematical symbols. Assume that the universe of discourse is the set of real numbers.

(a) There exists x such that $x^2 + 1$ is greater than zero.

(b) There exists a natural number n such that $n^2 = 36$.

(c) For every x, x^2 is greater than or equal to zero.

Problem 2.67. Express the formal definition of a limit (given above Definition 2.65) in logical and mathematical symbols.

If you look closely, many of the theorems that we have encountered up until this point were of the form $A(x) \implies B(x)$, where $A(x)$ and $B(x)$ are predicates. For example, consider Theorem 2.2, which states, "If n is an even integer, then n^2 is an even integer." In this case, "n is an even integer" and "n^2 is an even integer" are both predicates. So, it would be reasonable to assume that the entire theorem statement is a predicate. However, it is standard practice to interpret the sentence $A(x) \implies B(x)$ to mean $(\forall x)(A(x) \implies B(x))$ (where the universe of discourse for x needs to be made clear). We can also retool such statements to "hide" the implication. In particular, $(\forall x)(A(x) \implies B(x))$ has the same meaning as $(\forall x \in U')B(x)$, where U' is the collection of items from the universe of discourse U that makes $A(x)$ true. For example, we could rewrite the statement of Theorem 2.2 as "For every even integer n, n^2 is even."

Problem 2.68. Reword Theorem 2.7 so that it explicitly reads as a universally quantified statement. Compare with Problem 2.47.

Problem 2.69. Find at least two other instances of theorem statements that appeared earlier in the book and are written in the form $A(x) \implies B(x)$. Rewrite each in an equivalent way that makes the universal quantifier explicit while possibly suppressing the implication.

Problem 2.70. Consider the proposition "If $\varepsilon > 0$, then there exists $N \in \mathbb{N}$ such that $1/N < \varepsilon$." Assume the universe of discourse is the set $\mathbb{R}$.

(a) Express the statement in logical and mathematical symbols. Is the statement true?

(b) Reverse the order of the quantifiers to get a new statement. Does the meaning change? If so, how? Is the new statement true?

The symbolic expression $(\forall x)(\forall y)$ can be abbreviated as $\boxed{\forall x, y}$ as long as x and y are elements of the same universe.

Problem 2.71. Express the proposition "For all $x, y \in \mathbb{R}$ with $x < y$, there exists $m \in \mathbb{R}$ such that $x < m < y$" using logical and mathematical symbols.

Problem 2.72. Rewrite each of the following propositions in words and determine whether the proposition is true or false.

(a) $(\forall n \in \mathbb{N})(n^2 \geq 5)$

(b) $(\exists n \in \mathbb{N})(n^2 - 1 = 0)$

(c) $(\exists N \in \mathbb{N})(\forall n > N)(\frac{1}{n} < 0.01)$

(d) $(\forall m, n \in \mathbb{Z})((2|m \wedge 2|n) \implies 2|(m+n))$

(e) $(\forall x \in \mathbb{N})(\exists y \in \mathbb{N})(x - 2y = 0)$

(f) $(\exists x \in \mathbb{N})(\forall y \in \mathbb{N})(y \leq x)$

Problem 2.73. Consider the proposition $(\forall x)(\exists y)(xy = 1)$.

(a) Provide an example of a universe of discourse where this proposition is true.

(b) Provide an example of a universe of discourse where this proposition is false.

To whet your appetite for the next section, consider how you might prove a true proposition of the form "For all x...." If a proposition is false, then its negation is true. How would you go about negating a statement involving quantifiers?

> Like what you do, and then you will do your best.
>
> Katherine Johnson, mathematician

2.5 More About Quantification

When writing mathematical proofs, we do not explicitly use the symbolic representation of a given statement in terms of quantifiers and logical connectives. Nonetheless, having this notation at our disposal allows us to compartmentalize the abstract nature of mathematical propositions and provides us with a way to talk about the general structure involved in the construction of a proof.

Definition 2.74. Two quantified propositions are **logically equivalent** if they have the same truth value in every universe of discourse.

Problem 2.75. Consider the propositions $(\exists x \in U)(x^2 - 4 = 0)$ and $(\exists x \in U)(x^2 - 2 = 0)$, where U is some universe of discourse.

(a) Do these propositions have the same truth value if the universe of discourse is the set of real numbers?

(b) Provide an example of a universe of discourse such that the propositions yield different truth values.

(c) What can you conclude about the logical equivalence of these propositions?

It is worth pointing out an important distinction. Consider the propositions "All cars are red" and "All natural numbers are positive". Both of these are instances of the **logical form** $(\forall x)P(x)$. It turns out that the first proposition is false and the second is true; however, it does not make sense to attach a truth value to the logical form. A logical form is a blueprint for particular propositions. If we are careful, it makes sense to talk about whether two logical forms are logically equivalent. For example, $(\forall x)(P(x) \implies Q(x))$ is logically equivalent to $(\forall x)(\neg Q(x) \implies \neg P(x))$ since a conditional proposition is logically equivalent to its contrapositive (see Theorem 2.39). For fixed $P(x)$ and $Q(x)$, these two forms will always have the same truth value independent of the universe of discourse. If you change $P(x)$ and $Q(x)$, then the truth value may change, but the two forms will still agree.

The next theorem tells us how to negate logical forms involving quantifiers. Your proof should involve several mini arguments. For example, in Part (a), you will need to proof that if $\neg(\forall x)P(x)$ is true, then $(\exists x)(\neg P(x))$ is also true.

Theorem 2.76. *Let $P(x)$ be a predicate in some universe of discourse. Then*

(a) *$\neg(\forall x)P(x)$ is logically equivalent to $(\exists x)(\neg P(x))$;*

(b) *$\neg(\exists x)P(x)$ is logically equivalent to $(\forall x)(\neg P(x))$.*

Problem 2.77. Negate each of the following sentences. Disregard the truth value and the universe of discourse.

(a) $(\forall x)(x > 3)$

(b) $(\exists x)(x \text{ is prime} \wedge x \text{ is even})$

(c) All cars are red.

(d) Every Wookiee is named Chewbacca.

(e) Some hippies are Republican.

(f) Some birds are not angry.

(g) Not every video game will rot your brain.

(h) For all $x \in \mathbb{N}$, $x^2 + x + 41$ is prime.

(i) There exists $x \in \mathbb{Z}$ such that $1/x \notin \mathbb{Z}$.

(j) There is no function f such that if f is continuous, then f is not differentiable.

Using Theorem 2.76 and our previous results involving quantification, we can negate complex mathematical propositions by working from left to right. For example, if we negate the false proposition

$$(\exists x \in \mathbb{R})(\forall y \in \mathbb{R})(x + y = 0),$$

we obtain the proposition

$$\neg(\exists x \in \mathbb{R})(\forall y \in \mathbb{R})(x + y = 0),$$

which is logically equivalent to

$$(\forall x \in \mathbb{R})(\exists y \in \mathbb{R})(x + y \neq 0)$$

and must be true. For a more complicated example, consider the (false) proposition

$$(\forall x)[x > 0 \implies (\exists y)(y < 0 \wedge xy > 0)].$$

Then its negation

$$\neg(\forall x)[x > 0 \implies (\exists y)(y < 0 \wedge xy > 0)]$$

is logically equivalent to

$$(\exists x)[x > 0 \wedge \neg(\exists y)(y < 0 \wedge xy > 0)],$$

which happens to be logically equivalent to

$$(\exists x)[x > 0 \wedge (\forall y)(y \geq 0 \vee xy \leq 0)].$$

Can you identify the theorems that were used in the two examples above?

Problem 2.78. Negate each of the following propositions. Disregard the truth value and the universe of discourse.

(a) $(\forall n \in \mathbb{N})(\exists m \in \mathbb{N})(m < n)$

(b) For every $y \in \mathbb{R}$, there exists $x \in \mathbb{R}$ such that $y = x^2$.

(c) For all $y \in \mathbb{R}$, if y is not negative, then there exists $x \in \mathbb{R}$ such that $y = x^2$.

(d) For every $x \in \mathbb{R}$, there exists $y \in \mathbb{R}$ such that $y = x^2$.

(e) There exists $x \in \mathbb{R}$ such that for all $y \in \mathbb{R}$, $y = x^2$.

(f) There exists $y \in \mathbb{R}$ such that for all $x \in \mathbb{R}$, $y = x^2$.

(g) $(\forall x, y, z \in \mathbb{Z})((xy \text{ is even} \wedge yz \text{ is even}) \implies xz \text{ is even})$

(h) There exists a married person x such that for all married people y, x is married to y.

Problem 2.79. Consider the following proposition in some universe of discourse: For all goofy wobblers x, there exists a dinglehopper y such that if x is a not a nugget, then y is a doofus. Find the negation of this proposition so that it includes the phrase "is not a doofus."

Problem 2.80. Consider the following proposition in some universe of discourse: If x and y are both snazzy, then xy is not nifty. Find the contrapositive of this proposition so that it includes the phrase "not snazzy".

At this point, we should be able to use our understanding of quantification to construct counterexamples to complicated false propositions and proofs of complicated true propositions. Here are some general proof structures for various logical forms.

Skeleton Proof 2.81 (Direct Proof of $(\forall x)P(x)$). Here is the general structure for a direct proof of the proposition $(\forall x)P(x)$. Assume U is the universe of discourse.

Proof. [*State any upfront assumptions.*] Let $x \in U$.

... [*Use definitions and known results.*] ...

Therefore, $P(x)$ is true. Since x was arbitrary, for all x, $P(x)$. □

Combining Skeleton Proof 2.81 with Skeleton Proof 2.48, we obtain the following skeleton proof.

Skeleton Proof 2.82 (Proof of $(\forall x)(A(x) \implies B(x))$). Below is the general structure for a direct proof of the proposition $(\forall x)(A(x) \implies B(x)$. Assume U is the universe of discourse.

Proof. [*State any upfront assumptions.*] Let $x \in U$. Assume $A(x)$.

... [*Use definitions and known results to derive $B(x)$*] ...

Therefore, $B(x)$. □

Skeleton Proof 2.83 (Proof of $(\forall x)P(x)$ by Contradiction). Here is the general structure for a proof of the proposition $(\forall x)P(x)$ via contradiction. Assume U is the universe of discourse.

Proof. [*State any upfront assumptions.*] For sake of a contradiction, assume that there exists $x \in U$ such that $\neg P(x)$.

... [*Do something to derive a contradiction.*] ...

This is a contradiction. Therefore, for all x, $P(x)$ is true. □

Skeleton Proof 2.84 (Direct Proof of $(\exists x)P(x)$). Here is the general structure for a direct proof of the proposition $(\exists x)P(x)$. Assume U is the universe of discourse.

Proof. [*State any upfront assumptions.*] …

… [*Use definitions, axioms, and previous results to deduce that an x exists for which P(x) is true; or if you have an x that works, just verify that it does.*] …

Therefore, there exists $x \in U$ such that $P(x)$. □

Skeleton Proof 2.85 (Proof of $(\exists x)P(x)$ by Contradiction). Below is the general structure for a proof of the proposition $(\exists x)P(x)$ via contradiction. Assume U is the universe of discourse.

Proof. [*State any upfront assumptions.*] For sake of a contradiction, assume that for all $x \in U$, $\neg P(x)$.

… [*Do something to derive a contradiction.*] …

This is a contradiction. Therefore, there exists $x \in U$ such that $P(x)$. □

Note that if $Q(x)$ is a predicate for which $(\forall x)Q(x)$ is false, then a counterexample to this proposition amounts to showing $(\exists x)(\neg Q(x))$, which can be proved by following the structure of Skeleton Proof 2.84.

It is important to point out that sometimes we will have to combine various proof techniques in a single proof. For example, if you wanted to prove a proposition of the form $(\forall x)(P(x) \implies Q(x))$ by contradiction, we would start by assuming that there exists x in the universe of discourse such that $P(x)$ and $\neg Q(x)$.

Problem 2.86. Determine whether each of the following statements is true or false. If the statement is true, prove it. If the statement is false, provide a counterexample.

(a) For all $n \in \mathbb{N}$, $n^2 \geq 5$.

(b) There exists $n \in \mathbb{N}$ such that $n^2 - 1 = 0$.

(c) There exists $x \in \mathbb{N}$ such that for all $y \in \mathbb{N}$, $y \leq x$.

(d) For all $x \in \mathbb{Z}$, $x^3 \geq x$.

(e) For all $n \in \mathbb{Z}$, there exists $m \in \mathbb{Z}$ such that $n + m = 0$.

(f) There exists integers a and b such that $2a + 7b = 1$.

(g) There do not exist integers m and n such that $2m + 4n = 7$.

(h) For all $a, b, c \in \mathbb{Z}$, if a divides bc, then either a divides b or a divides c.

(i) For all $a, b \in \mathbb{Z}$, if ab is even, then either a or b is even.

Problem 2.87. Explain why the following "proof" is not a valid argument.

Claim. For all $x, y \in \mathbb{Z}$, if x and y are even, then $x + y$ is even.

"Proof." Suppose $x, y \in \mathbb{Z}$ such that x and y are even. For sake of a contradiction, assume that $x+y$ is odd. Then there exists $k \in \mathbb{Z}$ such that $x+y = 2k+1$. This implies that $(x + y) - 2k = 1$. We see that the left side of the equation is even because it is the difference of even numbers. However, the right side is odd. Since an even number cannot equal an odd number, we have a contradiction. Therefore, $x + y$ is even. □

Sometimes it is useful to split the universe of discourse into multiple collections to deal with separately. When doing this, it is important to make sure that your cases are exhaustive (i.e., every possible element of the universe of discourse has been accounted for). Ideally, your cases will also be disjoint (i.e., you have not considered the same element more than once). For example, if our universe of discourse is the set of integers, we can separately consider even versus odd integers. If our universe of discourse is the set of real numbers, we might want to consider rational versus irrational numbers, or possibly negative versus zero versus and positive. Attacking a proof in this way, is often referred to as a **proof by cases** (or **proof by exhaustion**). A proof by cases may also be useful when dealing with hypotheses involving "or". Note that the use of a proof by cases is justified by Theorem 2.30.

If you decide to approach a proof using cases, be sure to inform the reader that you are doing so and organize your proof in a sensible way. Note that doing an analysis of cases should be avoided if possible. For example, while it is valid to separately consider the cases of whether a is an even integer versus odd integer in the proof of Theorem 2.11, it is completely unnecessary. To prove the next theorem, you might want to consider two cases.

Theorem 2.88. *For all $n \in \mathbb{Z}$, $3n^2 + n + 14$ is even.*

Prove the following theorem by proving the contrapositive using two cases.

Theorem 2.89. *For all $n, m \in \mathbb{Z}$, if nm is odd, then n is odd and m is odd.*

When proving the previous theorem, you likely experienced some dèjá vu. You should have assumed "n is even or m is even" at some point in your proof. The first case is "n is even" while the second case is "m is even." (Note that you do not need to handle the case when both n and m are even since the two individual cases already yield the desired result.) The proofs for both cases are identical except the roles of n and m are interchanged. In instances such as this, mathematicians have a shortcut. Instead of writing two essentially identical proofs for each case, you can simply handle one of the cases and indicate that the remaining case follows from a nearly identical proof. The quickest way to do this is to use the phrase, "Without loss of generality, assume...". For example, here is a proof of Theorem 2.89 that utilizes this approach.

Proof of Theorem 2.89. We will prove the contrapositive. Let $n, m \in \mathbb{Z}$ and assume n is even or m is even. Without loss of generality, assume n is even. Then there exists $k \in \mathbb{Z}$ such that $n = 2k$. We see that

$$nm = (2k)m = 2(km).$$

Since both k and m are integers, km is an integer. This shows that nm is even. We have proved the contrapositive, and hence for all $n, m \in \mathbb{Z}$, if nm is odd, then n is odd and m is odd. □

Note that it would not be appropriate to utilize the "without loss of generality" approach to combine the two cases in the proof of Theorem 2.88 since the proof of the second case is not as simple as swapping the roles of symbols in the proof of the first case.

There are times when a theorem will make a claim about the *uniqueness* of a particular mathematical object. For example, in Section 5.1, you will be asked to prove that both the additive and multiplicative identities (i.e, 0 and 1) are unique (see Theorems 5.2 and 5.3). As another example, the Fundamental Theorem of Arithmetic (see Theorem 6.17) states that every natural number greater than 1 can be expressed uniquely (up to the order in which they appear) as the product of one or more primes. The typical approach to proving uniqueness is to suppose that there are potentially two objects with the desired property and then show that these objects are actually equal. Whether you approach this as a proof by contradiction is a matter of taste. It is common to use $\exists!$ as a symbolic abbreviation for "there exists a unique...such that".

Skeleton Proof 2.90 (Direct Proof of $(\exists!x)P(x)$). Here is the general structure for a direct proof of the proposition $(\exists!x)P(x)$. Assume U is the universe of discourse.

> *Proof.* [*State any upfront assumptions.*] ...
>
> ... [*Use definitions, axioms, and previous results to deduce that an x exists for which P(x) is true; or if you have an x that works, just verify that it does.*] ...

Therefore, there exists $x \in U$ such that $P(x)$. Now, suppose $x_1, x_2 \in U$ such that $P(x_1)$ and $P(x_2)$.

$$\ldots \text{ [\textit{Prove that} } x_1 = x_2.] \ldots$$

This implies that there exists a unique x such that $P(x)$. □

The next theorem provides an opportunity to practice proving uniqueness.

Theorem 2.91. *If $c, a, r \in \mathbb{R}$ such that $c \neq 0$ and $r \neq a/c$, then there exists a unique $x \in \mathbb{R}$ such that $(ax + 1)/(cx) = r$.*

> With two published novels and a file full of ideas for others, the only thing I know about writing is this: it only happens when you sit down and do it. Studying good writing is important, reading good writing is important, talking to other writers is important, but the only way you can produce good writing is to write.
>
> Jamie Beth Cohen, novelist

> Pass on what you have learned. Strength, mastery. But weakness, folly, failure also. Yes, failure most of all. The greatest teacher, failure is.
>
> Yoda, Jedi master

3

Set Theory

At its essence, all of mathematics is built on set theory. In this chapter, we will introduce some of the basics of sets and their properties.

3.1 Sets

Definition 3.1. A **set** is a collection of objects called **elements**. If A is a set and x is an element of A, we write $\boxed{x \in A}$. Otherwise, we write $\boxed{x \notin A}$. The set containing no elements is called the **empty set**, and is denoted by the symbol $\boxed{\emptyset}$. Any set that contains at least one element is referred to as a **nonempty set**.

If we think of a set as a box potentially containing some stuff, then the empty set is a box with nothing in it. One assumption we will make is that for any set A, $A \notin A$. The language associated to sets is specific. We will often define sets using the following notation, called **set-builder notation**:

$$\boxed{S = \{x \in A \mid P(x)\}},$$

where $P(x)$ is some predicate statement involving x. The first part "$x \in A$" denotes what type of x is being considered. The predicate to the right of the vertical bar (not to be confused with "divides") determines the condition(s) that each x must satisfy in order to be a member of the set. This notation is read as "The set of all x in A such that $P(x)$." As an example, the set $\{x \in \mathbb{N} \mid x \text{ is even and } x \geq 8\}$ describes the collection of even natural numbers that are greater than or equal to 8.

There are a few sets that are commonly discussed in mathematics and have predefined symbols to denote them. We have already encountered the integers, natural numbers, and real numbers. Notice that our definition of the rational numbers uses set-builder notation.

- **Natural Numbers:** $\mathbb{N} := \{1, 2, 3, \ldots\}$. Some books will include zero in the set of natural numbers, but we do not.
- **Integers:** $\mathbb{Z} := \{0, \pm 1, \pm 2, \pm 3, \ldots\}$.
- **Rational Numbers:** $\mathbb{Q} := \{a/b \mid a, b \in \mathbb{Z} \text{ and } b \neq 0\}$.
- **Real Numbers:** $\mathbb{R}$ denotes the set of real numbers. We are taking for granted that you have some familiarity with this set.

Since the set of natural numbers consists of the positive integers, the natural numbers are sometimes denoted by $\mathbb{Z}^+$.

Problem 3.2. Unpack the meaning of each of the following sets and provide a description of the elements that each set contains.

(a) $A = \{x \in \mathbb{N} \mid x = 3k \text{ for some } k \in \mathbb{N}\}$

(b) $B = \{t \in \mathbb{R} \mid t \leq 2 \text{ or } t \geq 7\}$

(c) $C = \{t \in \mathbb{Z} \mid t^2 \leq 2\}$

(d) $D = \{s \in \mathbb{Z} \mid -3 < s \leq 5\}$

(e) $E = \{m \in \mathbb{R} \mid m = 1 - \frac{1}{n}, \text{ where } n \in \mathbb{N}\}$

Problem 3.3. Write each of the following sentences using set-builder notation.

(a) The set of all real numbers less than $-\sqrt{2}$.

(b) The set of all real numbers greater than -12 and less than or equal to 42.

(c) The set of all even integers.

Parts (a) and (b) of Problem 3.3 are examples of intervals.

Definition 3.4. For $a, b \in \mathbb{R}$ with $a < b$, we define the following sets, referred to as **intervals**.

(a) $(a, b) := \{x \in \mathbb{R} \mid a < x < b\}$

(b) $[a, b] := \{x \in \mathbb{R} \mid a \leq x \leq b\}$

(c) $[a, b) := \{x \in \mathbb{R} \mid a \leq x < b\}$

(d) $(a, \infty) := \{x \in \mathbb{R} \mid a < x\}$

(e) $(-\infty, b) := \{x \in \mathbb{R} \mid x < b\}$

(f) $(-\infty, \infty) := \mathbb{R}$

We analogously define $(a, b]$, $[a, \infty)$, and $(-\infty, b]$. Intervals of the form (a, b), $(-\infty, b)$, (a, ∞), and $(-\infty, \infty)$ are called **open intervals** while $[a, b]$ is referred to as a **closed interval**. A **bounded interval** is any interval of the form (a, b), $[a, b)$, $(a, b]$, and $[a, b]$. For bounded intervals, a and b are called the **endpoints** of the interval.

We will always assume that any time we write $(a, b), [a, b], (a, b]$, or $[a, b)$ that $a < b$. We will see where the terminology of "open" and "closed" comes from in Section 5.2.

Problem 3.5. Give an example of each of the following.

(a) An interval that is neither an open nor closed interval.

(b) An infinite set that is not an interval.

Definition 3.6. If A and B are sets, then we say that A is a **subset** of B, written $A \subseteq B$, provided that every element of A is an element of B.

Problem 3.7. List all of the subsets of $A = \{1, 2, 3\}$.

Every nonempty set always has two subsets. Notice that if $A = \emptyset$, then Parts (a) and (b) of the next theorem say the same thing.

Theorem 3.8. *Let A be a set. Then*

(a) $A \subseteq A$, *and*

(b) $\emptyset \subseteq A$.

Observe that "$A \subseteq B$" is equivalent to "For all x (in the universe of discourse), if $x \in A$, then $x \in B$." Since we know how to deal with "for all" statements and conditional propositions, we know how to go about proving $A \subseteq B$. If A happens to be the empty set, then the statement "For all x (in the universe of discourse), if $x \in A$, then $x \in B$" is vacuously true. This is in agreement with Theorem 3.8(b), which states that the empty set is always a subset of every set. In light of this, it is common to omit discussion of the case when A is the empty set when proving that A is s a subset of B.

Problem 3.9. Suppose A and B are sets. Describe a skeleton proof for proving that $A \subseteq B$.

Theorem 3.10 (Transitivity of Subsets)**.** *Suppose that A, B, and C are sets. If $A \subseteq B$ and $B \subseteq C$, then $A \subseteq C$.*

Definition 3.11. Two sets A and B are **equal**, denoted $\boxed{A = B}$, if the sets contain the same elements.

Since the next theorem is a biconditional proposition, you need to write two distinct subproofs, one for "$A = B$ implies $A \subseteq B$ and $B \subseteq A$", and another for "$A \subseteq B$ and $B \subseteq A$ implies $A = B$". Be sure to make it clear to the reader when you are proving each implication.

Theorem 3.12. *Suppose that A and B are sets. Then $A = B$ if and only if $A \subseteq B$ and $B \subseteq A$.*

Note that if we want to prove $A = B$, then we have to do two separate subproofs: one for $A \subseteq B$ and one for $B \subseteq A$. Be sure to make it clear to the reader where these subproofs begin and end. One approach is to label each subproof with "($\subseteq$)" and "($\supseteq$)" (including the parentheses), respectively.

Definition 3.13. If $A \subseteq B$, then A is called a **proper subset** provided that $A \neq B$. In this case, we may write $\boxed{A \subset B}$ or $\boxed{A \subsetneq B}$.

Note that some authors use $\subset$ to mean $\subseteq$, so some confusion could arise if you are not reading carefully.

Definition 3.14. Let A and B be sets in some universe of discourse U.

(a) The **union** of the sets A and B is $\boxed{A \cup B := \{x \in U \mid x \in A \text{ or } x \in B\}}$.

(b) The **intersection** of the sets A and B is $\boxed{A \cap B := \{x \in U \mid x \in A \text{ and } x \in B\}}$.

(c) The **set difference** of the sets A and B is $\boxed{A \setminus B := \{x \in U \mid x \in A \text{ and } x \notin B\}}$.

(d) The **complement of** A (relative to U) is the set

$$\boxed{A^c := U \setminus A = \{x \in U \mid x \notin A\}}.$$

Definition 3.15. If two sets A and B have the property that $A \cap B = \emptyset$, then we say that A and B are **disjoint** sets.

Problem 3.16. Suppose that the universe of discourse is $U = \{1, 2, 3, 4, 5, 6, 7, 8, 9, 10\}$. Let $A = \{1, 2, 3, 4, 5\}$, $B = \{1, 3, 5\}$, and $C = \{2, 4, 6, 8\}$. Find each of the following.

(a) $A \cap C$

(b) $B \cap C$

(c) $A \cup B$

(d) $A \setminus B$

(e) $B \setminus A$

(f) $C \setminus B$

(g) B^c

(h) A^c

(i) $(A \cup B)^c$

(j) $A^c \cap B^c$

Problem 3.17. Suppose that the universe of discourse is $U = \mathbb{R}$. Let $A = [-3, -1)$, $B = (-2.5, 2)$, and $C = (-2, 0]$. Find each of the following.

(a) A^c

(b) $A \cap C$

(c) $A \cap B$

(d) $A \cup B$

(e) $(A \cap B)^c$

(f) $(A \cup B)^c$

(g) $A \setminus B$

(h) $A \setminus (B \cup C)$

(i) $B \setminus A$

Problem 3.18. Suppose that the universe of discourse is $U = \{x, y, z, \{y\}, \{x, z\}\}$. Let $S = \{x, y, z\}$ and $T = \{x, \{y\}\}$. Find each of the following.

(a) $S \cap T$

(b) $(S \cup T)^c$

(c) $T \setminus S$

Theorem 3.19. *If A and B are sets such that $A \subseteq B$, then $B^c \subseteq A^c$.*

Theorem 3.20. *If A and B are sets, then $A \setminus B = A \cap B^c$.*

In Chapter 2, we encountered De Morgan's Law (see Theorem 2.26 and Problem 2.27), which provided a method for negating compound propositions involving conjunctions and disjunctions. The next theorem provides a method for taking the complement of unions and intersections of sets. This result is also known as De Morgan's Law. Do you see why?

Theorem 3.21 (De Morgan's Law). *If A and B are sets, then*

(a) $(A \cup B)^c = A^c \cap B^c$, *and*

(b) $(A \cap B)^c = A^c \cup B^c$.

The next theorem indicates how intersections and unions interact with each other.

Theorem 3.22 (Distribution of Union and Intersection). *If A, B, and C are sets, then*

(a) $A \cup (B \cap C) = (A \cup B) \cap (A \cup C)$, *and*

(b) $A \cap (B \cup C) = (A \cap B) \cup (A \cap C)$.

Problem 3.23. For each of the statements (a)–(d) on the left, find an equivalent symbolic proposition chosen from the list (i)–(v) on the right. Note that not every statement on the right will get used.

(a) $A \not\subseteq B$.

(b) $A \cap B = \emptyset$.

(c) $(A \cup B)^c \neq \emptyset$.

(d) $(A \cap B)^c = \emptyset$.

(i) $(\forall x)(x \in A \wedge x \in B)$

(ii) $(\forall x)(x \in A \implies x \notin B)$

(iii) $(\exists x)(x \notin A \wedge x \notin B)$

(iv) $(\exists x)(x \in A \vee x \in B)$

(v) $(\exists x)(x \in A \wedge x \notin B)$

> In mathematics the art of proposing a question must be held of higher value than solving it.
>
> Georg Cantor, mathematician

3.2 Russell's Paradox

We now turn our attention to the issue of whether there is one mother of all universal sets. Before reading any further, consider this for a moment. That is, is there one largest set that all other sets are a subset of? Or, in other words, is there a set of all sets? To help wrap our heads around this issue, consider the following riddle, known as the **Barber of Seville Paradox**.

> In Seville, there is a barber who shaves all those men, and only those men, who do not shave themselves. Who shaves the barber?

Problem 3.24. In the Barber of Seville Paradox, does the barber shave himself or not?

Problem 3.24 is an example of a **paradox**. A paradox is a statement that can be shown, using a given set of axioms and definitions, to be both true and false.

Recall that an axiom is a statement that is assumed to be true without proof. These are the basic building blocks from which all theorems are proved. Paradoxes are often used to show the inconsistencies in a flawed axiomatic theory. The term paradox is also used informally to describe a surprising or counterintuitive result that follows from a given set of rules. Now, suppose that there is a set of all sets and call it $\mathcal{U}$. That is, $\mathcal{U} := \{A \mid A \text{ is a set}\}$.

Problem 3.25. Given our definition of $\mathcal{U}$, explain why $\mathcal{U}$ is an element of itself.

If we continue with this line of reasoning, it must be the case that some sets are elements of themselves and some are not. Let X be the set of all sets that are elements of themselves and let Y be the set of all sets that are not elements of themselves.

Problem 3.26. Does Y belong to X or Y? Explain why this is a paradox.

The above paradox is one way of phrasing a paradox referred to as **Russell's Paradox**, named after British mathematician and philosopher Bertrand Russell (1872–1970). How did we get into this mess in the first place?! By assuming the existence of a set of all sets, we can produce all sorts of paradoxes. The only way to avoid these types of paradoxes is to conclude that there is no set of all sets. That is, the collection of all sets cannot be a set itself.

According to naive set theory (i.e., approaching set theory using natural language as opposed to formal logic), any definable collection is a set. As Russell's Paradox illustrates, this leads to problems. It turns out that any proposition can be proved from a contradiction, and hence the presence of contradictions like Russell's Paradox would appear to be catastrophic for mathematics. Since set theory is often viewed as the basis for axiomatic development in mathematics, Russell's Paradox calls the foundations of mathematics into question. In response to this threat, a great deal of research went into developing consistent axioms (i.e., free of contradictions) for set theory in the early 20th century. In 1908, Ernst Zermelo (1871–1953) proposed a collection of axioms for set theory that avoided the inconsistencies of naive set theory. In the 1920s, adjustments to Zermelo's axioms were made by Abraham Fraenkel (1891–1965), Thoralf Skolem (1887–1963), and Zermelo that resulted in a collection of nine axioms, called ZFC, where ZF stands for Zermelo and Fraenkel and C stands for the Axiom of Choice, which is one of the nine axioms. Loosely speaking, the Axiom of Choice states that given any collection of sets, each containing at least one element, it is possible to make a selection of exactly one object from each set, even if the collection of sets is infinite. There was a period of time in mathematics when the Axiom of Choice was controversial, but nowadays it is generally accepted. There is a fascinating history concerning the Axiom of Choice, including its controversy. The Wikipedia page for the Axiom of Choice is a good place to start if you are interested in learning more. There are several competing axiomatic approaches to set theory, but ZFC is considered the canonical collection of axioms by most mathematicians.

Appendix C includes a few more examples of paradoxes, which you are encouraged to ponder.

> In times of change, learners inherit the earth, while the learned find themselves beautifully equipped to deal with a world that no longer exists.
>
> Eric Hoffer, moral and social philosopher

3.3 Power Sets

We have already seen that using union, intersection, set difference, and complement we can create new sets (in the same universe) from existing sets. In this section, we will describe another way to generate new sets; however, the new sets will not "live" in the same universe this time. The following set is always a set of subsets. That is, its elements are themselves sets.

Definition 3.27. If S is a set, then the **power set** of S is the set of subsets of S. The power set of S is denoted $\boxed{\mathcal{P}(S)}$.

You can see that a power set of S is not composed of *elements* of S, but rather it is composed of *subsets* of S, and none of these subsets are elements of S.

For example, if $S = \{a, b\}$, then $\mathcal{P}(S) = \{\emptyset, \{a\}, \{b\}, S\}$. It follows immediately from the definition that $A \subseteq S$ if and only if $A \in \mathcal{P}(S)$.

Problem 3.28. For each of the following sets, find the power set.

(a) $A = \{\circ, \triangle, \square\}$

(b) $B = \{a, \{a\}\}$

(c) $C = \emptyset$

(d) $D = \{\emptyset\}$

Problem 3.29. How many subsets do you think that a set with n elements has? What if $n = 0$? You do not need to prove your conjecture at this time. We will prove this later using mathematical induction.

It is important to realize that the concepts of *element* and *subset* need to be carefully delineated. For example, consider the set $A = \{x, y\}$. The object x is an element of A, but the object $\{x\}$ is both a subset of A and an element of $\mathcal{P}(A)$. This can get confusing rather quickly. Consider the set B from Problem 3.28. The set

$\{a\}$ happens to be an element of B, a subset of B, and an element of $\mathcal{P}(B)$. The upshot is that it is important to pay close attention to whether "$\subseteq$" or "$\in$" is the proper symbol to use.

Since the next theorem is a biconditional proposition, you need to write two distinct subproofs, one for "$S \subseteq T$ implies $\mathcal{P}(S) \subseteq \mathcal{P}(T)$", and another for "$\mathcal{P}(S) \subseteq \mathcal{P}(T)$ implies $S \subseteq T$".

Theorem 3.30. *Let S and T be sets. Then $S \subseteq T$ if and only if $\mathcal{P}(S) \subseteq \mathcal{P}(T)$.*

Problem 3.31. Let S and T be sets. Determine whether each of the following statements is true or false. If the statement is true, prove it. If the statement is false, provide a counterexample.

(a) $\mathcal{P}(S \cap T) \subseteq \mathcal{P}(S) \cap \mathcal{P}(T)$

(b) $\mathcal{P}(S) \cap \mathcal{P}(T) \subseteq \mathcal{P}(S \cap T)$

(c) $\mathcal{P}(S \cup T) \subseteq \mathcal{P}(S) \cup \mathcal{P}(T)$

(d) $\mathcal{P}(S) \cup \mathcal{P}(T) \subseteq \mathcal{P}(S \cup T)$

While power sets provide a useful way of generating new sets, they also play a key role in Georg Cantor's (1845–1918) investigation into the "size" of sets. **Cantor's Theorem** (see Theorem 9.64) states that the power set of a set—even if the set is infinite—is always "larger" than the original set. One consequence of this is that there are different sizes of infinity and no largest infinity. Mathematics is awesome.

> The master has failed more times than the beginner has even tried.
>
> Stephen McCranie, author & illustrator

3.4 Indexing Sets

Suppose we consider the following collection of open intervals:

$$(0, 1), (0, 1/2), (0, 1/4), \ldots, (0, 1/2^{n-1}), \ldots$$

This collection has a natural way for us to "index" the sets:

$$I_1 = (0, 1), I_2 = (0, 1/2), \ldots, I_n = (0, 1/2^{n-1}), \ldots$$

In this case the sets are **indexed** by the set $\mathbb{N}$. The subscripts are taken from the **index set**. If we wanted to talk about an arbitrary set from this indexed collection, we could use the notation I_n.

Let's consider another example:

$$\{a\}, \{a,b\}, \{a,b,c\}, \dots, \{a,b,c,\dots,z\}$$

An obvious way to index these sets is as follows:

$$A_1 = \{a\}, A_2 = \{a,b\}, A_3 = \{a,b,c\}, \dots, A_{26} = \{a,b,c,\dots,z\}$$

In this case, the collection of sets is indexed by $\{1, 2, \dots, 26\}$.

Using indexing sets in mathematics is an extremely useful notational tool, but it is important to keep straight the difference between the sets that are being indexed, the elements in each set being indexed, the indexing set, and the elements of the indexing set.

Any set (finite or infinite) can be used as an indexing set. Often capital Greek letters are used to denote arbitrary indexing sets and small Greek letters to represent elements of these sets. If the indexing set is a subset of $\mathbb{R}$, then it is common to use Roman letters as individual indices. Of course, these are merely conventions, not rules.

- If Δ is a set and we have a collection of sets indexed by Δ, then we may write $\{S_\alpha\}_{\alpha\in\Delta}$ to refer to this collection. We read this as "the set of S-sub-alphas over alpha in Delta."
- If a collection of sets is indexed by $\mathbb{N}$, then we may write $\{U_n\}_{n\in\mathbb{N}}$ or $\{U_n\}_{n=1}^{\infty}$.
- Borrowing from this idea, a collection $\{A_1, \dots, A_{26}\}$ may be written as $\{A_n\}_{n=1}^{26}$.

Definition 3.32. Let $\{A_\alpha\}_{\alpha\in\Delta}$ be a collection of sets.

(a) The **union of the entire collection** is defined via

$$\bigcup_{\alpha\in\Delta} A_\alpha := \{x \mid x \in A_\alpha \text{ for some } \alpha \in \Delta\}.$$

(b) The **intersection of the entire collection** is defined via

$$\bigcap_{\alpha\in\Delta} A_\alpha := \{x \mid x \in A_\alpha \text{ for all } \alpha \in \Delta\}.$$

In the special case that $\Delta = \mathbb{N}$, we write

$$\bigcup_{n=1}^{\infty} A_n = \{x \mid x \in A_n \text{ for some } n \in \mathbb{N}\} = A_1 \cup A_2 \cup A_3 \cup \cdots$$

and

$$\bigcap_{n=1}^{\infty} A_n = \{x \mid x \in A_n \text{ for all } n \in \mathbb{N}\} = A_1 \cap A_2 \cap A_3 \cap \cdots$$

Similarly, if $\Delta = \{1, 2, 3, 4\}$, then

$$\bigcup_{n=1}^{4} A_n = A_1 \cup A_2 \cup A_3 \cup A_4$$

and

$$\bigcap_{n=1}^{4} A_n = A_1 \cap A_2 \cap A_3 \cap A_4.$$

Notice the difference between "$\bigcup$" and "$\cup$" (respectively, "$\bigcap$" and "$\cap$").

Problem 3.33. Let $\{I_n\}_{n\in\mathbb{N}}$ be the collection of open intervals from the beginning of the section. Find each of the following.

(a) $\displaystyle\bigcup_{n\in\mathbb{N}} I_n$

(b) $\displaystyle\bigcap_{n\in\mathbb{N}} I_n$

Problem 3.34. Let $\{A_n\}_{n=1}^{26}$ be the collection from earlier in the section. Find each of the following.

(a) $\displaystyle\bigcup_{n=1}^{26} A_n$

(b) $\displaystyle\bigcap_{n=1}^{26} A_n$

Problem 3.35. Let $S_n = \{x \in \mathbb{R} \mid n-1 < x < n\}$, where $n \in \mathbb{N}$. Find each of the following.

(a) $\displaystyle\bigcup_{n=1}^{\infty} S_n$

(b) $\displaystyle\bigcap_{n=1}^{\infty} S_n$

Problem 3.36. Let $T_n = \{x \in \mathbb{R} \mid -\frac{1}{n} < x < \frac{1}{n}\}$, where $n \in \mathbb{N}$. Find each of the following.

(a) $\displaystyle\bigcup_{n=1}^{\infty} T_n$

(b) $\displaystyle\bigcap_{n=1}^{\infty} T_n$

Problem 3.37. For each $r \in \mathbb{Q}$ (the rational numbers), let N_r be the set containing all real numbers *except* r. Find each of the following.

(a) $\displaystyle\bigcup_{r \in \mathbb{Q}} N_r$

(b) $\displaystyle\bigcap_{r \in \mathbb{Q}} N_r$

Definition 3.38. A collection of sets $\{A_\alpha\}_{\alpha \in \Delta}$ is **pairwise disjoint** if $A_\alpha \cap A_\beta = \emptyset$ for $\alpha \neq \beta$.

Problem 3.39. Provide an example of a collection of sets $\{A_\alpha\}_{\alpha \in \Delta}$ that is not pairwise disjoint even though $\bigcap_{\alpha \in \Delta} A_\alpha = \emptyset$.

Problem 3.40. For each of the following, provide an example of a collection of sets with the stated property.

(a) A collection of three subsets of $\mathbb{R}$ such that the collection is not pairwise disjoint, the union equals $\mathbb{R}$, and the intersection of the collection is empty.

(b) A collection of infinitely many subsets of $\mathbb{R}$ such that the collection is not pairwise disjoint, the union equals $\mathbb{R}$, and the intersection of the collection is empty.

(c) A collection of infinitely many subsets of $\mathbb{R}$ such that the collection is pairwise disjoint, the union equals $\mathbb{R}$, and the intersection of the collection is empty.

Theorem 3.41 (Generalized Distribution of Union and Intersection). *Let $\{A_\alpha\}_{\alpha \in \Delta}$ be a collection of sets and let B be any set. Then*

(a) $\displaystyle B \cup \left(\bigcap_{\alpha \in \Delta} A_\alpha \right) = \bigcap_{\alpha \in \Delta} (B \cup A_\alpha)$, *and*

(b) $\displaystyle B \cap \left(\bigcup_{\alpha \in \Delta} A_\alpha \right) = \bigcup_{\alpha \in \Delta} (B \cap A_\alpha)$.

Theorem 3.42 (Generalized De Morgan's Law). *Let $\{A_\alpha\}_{\alpha \in \Delta}$ be a collection of sets. Then*

(a) $\displaystyle \left(\bigcup_{\alpha \in \Delta} A_\alpha \right)^C = \bigcap_{\alpha \in \Delta} A_\alpha^C$, *and*

(b) $\displaystyle \left(\bigcap_{\alpha \in \Delta} A_\alpha \right)^C = \bigcup_{\alpha \in \Delta} A_\alpha^C$.

At the end of Section 3.2, we mentioned the Axiom of Choice. Using the language of indexing sets, we can now state this axiom precisely.

Axiom 3.43 (Axiom of Choice). *For every indexed collection $\{A_\alpha\}_{\alpha\in\Delta}$ of non-empty sets, there exists an indexed collection $\{a_\alpha\}_{\alpha\in\Delta}$ of elements such that $a_\alpha \in A_\alpha$ for each $\alpha \in \Delta$.*

Intuitively, the Axiom of Choice guarantees the existence of mathematical objects that are obtained by a sequence of choices. It applies to both the finite and infinite setting. As an analogy, we can think of each A_α as a drawer in a dresser and each a_α as an article of clothing chosen from the drawer identified with A_α. The Axiom of Choice is surprisingly powerful, sometimes leading to unexpected consequences. It often gets used in subtle ways that mathematicians are not always explicit with. We will require the Axiom of Choice when proving Theorems 9.31 and 9.47. When proving these theorems, be on the lookout for where you are invoking the Axiom of Choice.

> All sorts of things can happen when you're open to new ideas and playing around with things.
>
> ---
>
> Stephanie Kwolek, chemist

3.5 Cartesian Products of Sets

Given a collection of sets, we can form new sets by taking unions, intersections, complements, and set differences. In this section, we introduce a type of "product" of sets. You have already encountered this concept when you learned to plot points in the plane. You also crossed paths with this notion if you have taken a course in linear algebra.

Definition 3.44. For each $n \in \mathbb{N}$, we define an n-**tuple** to be an ordered list of n elements of the form $\boxed{(a_1, a_2, \ldots, a_n)}$. We refer to a_i as the ith **component** (or **coordinate**) of $(a_1, a_2, \ldots, a_n)$. Two n-tuples $(a_1, a_2, \ldots, a_n)$ and $(b_1, b_2, \ldots, b_n)$ are equal if $a_i = b_i$ for all $1 \leq i \leq n$. A 2-tuple (a, b) is more commonly referred to as an **ordered pair** while a 3-tuple (a, b, c) is often called an **ordered triple**.

Occasionally, other symbols are used to surround the components of an n-tuple, such as square brackets "[ ]" or angle brackets "$\langle\ \rangle$". In some programming languages, curly braces "{ }" are used to specify arrays. However, we avoid this convention in mathematics since curly braces are the standard notation for sets. The term "tuple" can also occur when discussing other mathematical objects, such as vectors.

We can use the notion of n-tuples to construct new sets from existing sets.

Definition 3.45. If A and B are sets, the **Cartesian product** (or **direct product**) of A and B, denoted $A \times B$ (read as "A times B" or "A cross B"), is the set of all ordered pairs where the first component is from A and the second component is from B. In set-builder notation, we have

$$\boxed{A \times B := \{(a, b) \mid a \in A, b \in B\}}.$$

We similarly define the Cartesian product of n sets, say $A_1, \dots, A_n$, by

$$\boxed{\prod_{i=1}^{n} A_i := A_1 \times \cdots \times A_n := \{(a_1, \dots, a_n) \mid a_j \in A_j \text{ for all } 1 \leq j \leq n\}},$$

where A_i is referred to as the ith **factor** of the Cartesian product. As a special case, the set

$$\underbrace{A \times \cdots \times A}_{n \text{ factors}}$$

is often abbreviated as A^n.

Cartesian products are named after French philosopher and mathematician René Descartes (1596–1650). Cartesian products will play a prominent role in Chapter 7.

Example 3.46. If $A = \{a, b, c\}$ and $B = \{☺, ☹\}$, then

$$A \times B = \{(a, ☺), (a, ☹), (b, ☺), (b, ☹), (c, ☺), (c, ☹)\}.$$

Example 3.47. The standard two-dimensional plane $\mathbb{R}^2$ and standard three space $\mathbb{R}^3$ are familiar examples of Cartesian products. In particular, we have

$$\mathbb{R}^2 = \mathbb{R} \times \mathbb{R} = \{(x, y) \mid x, y \in \mathbb{R}\}$$

and

$$\mathbb{R}^3 = \mathbb{R} \times \mathbb{R} \times \mathbb{R} = \{(x, y, z) \mid x, y, z \in \mathbb{R}\}.$$

Problem 3.48. Consider the sets A and B from Example 3.46.

(a) Find $B \times A$.

(b) Find $B \times B$.

Problem 3.49. If A and B are sets, why do you think that $A \times B$ is referred to as a type of "product"? Think about the area model for multiplication of natural numbers.

Problem 3.50. If A and B are both finite sets, then how many elements will $A \times B$ have?

Problem 3.51. Let $A = \{1, 2, 3\}$, $B = \{1, 2\}$, and $C = \{1, 3\}$. Find $A \times B \times C$.

Problem 3.52. Let $X = [0, 1]$ and $Y = \{1\}$. Write each of the following using set-builder notation and then describe the set geometrically (e.g., draw a picture).

(a) $X \times Y$

(b) $Y \times X$

(c) $X \times X$

(d) $Y \times Y$

Problem 3.53. If A is a set, then what is $A \times \emptyset$ equal to?

Problem 3.54. Given sets A and B, when will $A \times B$ be equal to $B \times A$?

Problem 3.55. Write $\mathbb{N} \times \mathbb{R}$ using set-builder notation and then describe this set geometrically by interpreting it as a subset of $\mathbb{R}^2$.

We now turn our attention to subsets of Cartesian products.

Theorem 3.56. *Let A, B, C, and D be sets. If $A \subseteq C$ and $B \subseteq D$, then $A \times B \subseteq C \times D$.*

Problem 3.57. Is it true that if $A \times B \subseteq C \times D$, then $A \subseteq C$ and $B \subseteq D$? Do not forget to think about cases involving the empty set.

Problem 3.58. Is every subset of $C \times D$ of the form $A \times B$, where $A \subseteq C$ and $B \subseteq D$? If so, prove it. If not, find a counterexample.

Problem 3.59. If A, B, and C are nonempty sets, is $A \times B$ a subset of $A \times B \times C$?

Problem 3.60. Let $A = [2, 5]$, $B = [3, 7]$, $C = [1, 3]$, and $D = [2, 4]$. Compute each of the following.

(a) $(A \cap B) \times (C \cap D)$

(b) $(A \times C) \cap (B \times D)$

(c) $(A \cup B) \times (C \cup D)$

(d) $(A \times C) \cup (B \times D)$

(e) $A \times (B \cap C)$

(f) $(A \times B) \cap (A \times C)$

(g) $A \times (B \cup C)$

(h) $(A \times B) \cup (A \times C)$

Problem 3.61. Let A, B, C, and D be sets. Determine whether each of the following statements is true or false. If a statement is true, prove it. Otherwise, provide a counterexample.

(a) $(A \cap B) \times (C \cap D) = (A \times C) \cap (B \times D)$

(b) $(A \cup B) \times (C \cup D) = (A \times C) \cup (B \times D)$

(c) $A \times (B \cap C) = (A \times B) \cap (A \times C)$

(d) $A \times (B \cup C) = (A \times B) \cup (A \times C)$

(e) $A \times (B \setminus C) = (A \times B) \setminus (A \times C)$

Problem 3.62. If A and B are sets, conjecture a way to rewrite $(A \times B)^C$ in a way that involves A^C and B^C and then prove your conjecture.

If there is no struggle, there is no progress.

Frederick Douglass, writer & statesman

> Every time that a human being succeeds in making an effort of attention with the sole idea of increasing [their] grasp of truth, [they acquire] a greater aptitude for grasping it, even if [their] effort produces no visible fruit.
>
> Simone Weil, philosopher & political activist

4 Induction

In this chapter, we introduce mathematical induction, which is a proof technique that is useful for proving statements of the form $(\forall n \in \mathbb{N})P(n)$, or more generally $(\forall n \in \mathbb{Z})(n \geq a \implies P(n))$, where $P(n)$ is some predicate and $a \in \mathbb{Z}$.

4.1 Introduction to Induction

Consider the claims:

(a) For all $n \in \mathbb{N}$, $1 + 2 + 3 + \cdots + n = \dfrac{n(n+1)}{2}$.

(b) For all $n \in \mathbb{N}$, $n^2 + n + 41$ is prime.

Let's take a look at potential proofs.

"Proof" of (a). If $n = 1$, then $1 = \frac{1(1+1)}{2}$. If $n = 2$, then $1 + 2 = 3 = \frac{2(2+1)}{2}$. If $n = 3$, then $1 + 2 + 3 = 6 = \frac{3(3+1)}{2}$, and so on. □

"Proof" of (b). If $n = 1$, then $n^2 + n + 41 = 43$, which is prime. If $n = 2$, then $n^2 + n + 41 = 47$, which is prime. If $n = 3$, then $n^2 + n + 41 = 53$, which is prime, and so on. □

Are these actual proofs? No! In fact, the second claim isn't even true. If $n = 41$, then $n^2 + n + 41 = 41^2 + 41 + 41 = 41(41 + 1 + 1)$, which is not prime since it has 41 as a factor. It turns out that the first claim is true, but what we wrote cannot be a proof since the same type of reasoning when applied to the second claim seems to prove something that isn not actually true. We need a rigorous way of capturing "and so on" and a way to verify whether it really is "and so on."

Recall that an axiom is a basic mathematical assumption. The following axiom is one of the Peano Axioms, which is a collection of axioms for the natural numbers introduced in the 19th century by Italian mathematician Giuseppe Peano (1858–1932).

Axiom 4.1 (Axiom of Induction). *Let $S \subseteq \mathbb{N}$ such that both*

(i) $1 \in S$, *and*

(ii) *if* $k \in S$, *then* $k + 1 \in S$.

Then $S = \mathbb{N}$.

We can think of the set S as a ladder, where the first hypothesis as saying that we have a first rung of a ladder. The second hypothesis says that if we are on any arbitrary rung of the ladder, then we can always get to the next rung. Taken together, this says that we can get from the first rung to the second, from the second to the third, and in general, from any kth rung to the $(k + 1)$st rung, so that our ladder is actually $\mathbb{N}$. Do you agree that the Axiom of Induction is a pretty reasonable assumption?

At the end of Section 3.2, we briefly discussed ZFC, which is the standard choice for axiomatic set theory. It turns out that one can prove the Axiom of Induction as a theorem in ZFC. However, that will not be the approach we take. Instead, we are assuming the Axiom of Induction is true. Using this axiom, we can prove the following theorem, known as the **Principle of Mathematical Induction**. One approach to proving this theorem is to let $S = \{k \in \mathbb{N} \mid P(k) \text{ is true}\}$ and use the Axiom of Induction. The set S is sometimes called the **truth set**. Your job is to show that the truth set is all of $\mathbb{N}$.

Theorem 4.2 (Principle of Mathematical Induction). *Let* $P(1), P(2), P(3), \dots$ *be a sequence of statements, one for each natural number. Assume*

(i) $P(1)$ *is true, and*

(ii) *if* $P(k)$ *is true, then* $P(k + 1)$ *is true.*

Then $P(n)$ *is true for all* $n \in \mathbb{N}$.

The Principle of Mathematical Induction provides us with a process for proving statements of the form: "For all $n \in \mathbb{N}$, $P(n)$," where $P(n)$ is some predicate involving n. Hypothesis (i) above is called the **base step** (or **base case**) while (ii) is called the **inductive step**.

You should not confuse *mathematical induction* with *inductive reasoning* associated with the natural sciences. Inductive reasoning is a scientific method whereby one induces general principles from observations. On the other hand, mathematical induction is a deductive form of reasoning used to establish the validity of a proposition.

Skeleton Proof 4.3 (Proof of $(\forall n \in \mathbb{N})P(n)$ by Induction). Here is the general structure for a proof by induction.

Proof. We proceed by induction.

(i) Base step: [*Verify that $P(1)$ is true. This often, but not always, amounts to plugging $n = 1$ into two sides of some claimed equation and verifying that both sides are actually equal.*]

(ii) Inductive step: [*Your goal is to prove "For all $k \in \mathbb{N}$, if $P(k)$ is true, then $P(k+1)$ is true."*] Let $k \in \mathbb{N}$ and assume that $P(k)$ is true. [*Do something to derive that $P(k+1)$ is true.*] Therefore, $P(k+1)$ is true.

Thus, by induction, $P(n)$ is true for all $n \in \mathbb{N}$. □

Prove the next few theorems using induction. The first result may look familiar from calculus. Recall that $\sum_{i=1}^{n} i = 1 + 2 + 3 + \cdots + n$, by definition.

Theorem 4.4. *For all $n \in \mathbb{N}$, $\sum_{i=1}^{n} i = \frac{n(n+1)}{2}$.*

Theorem 4.5. *For all $n \in \mathbb{N}$, 3 divides $4^n - 1$.*

Theorem 4.6. *For all $n \in \mathbb{N}$, 6 divides $n^3 - n$.*

Theorem 4.7. *Let $p_1, p_2, \ldots, p_n$ be n distinct points arranged on a circle. Then the number of line segments joining all pairs of points is $\frac{n^2-n}{2}$.*

Problem 4.8. Consider a grid of squares that is 2^n squares wide by 2^n squares long, where $n \in \mathbb{N}$. One of the squares has been cut out, but you do not know which one! You have a bunch of L-shapes made up of 3 squares. Prove that you can perfectly cover this chessboard with the L-shapes (with no overlap) for any $n \in \mathbb{N}$. Figure 4.1 depicts one possible covering for the case involving $n = 2$.

> Do not stop thinking of life as an adventure. You have no security unless you can live bravely, excitingly, imaginatively; unless you can choose a challenge instead of competence.
>
> Eleanor Roosevelt, political figure & activist

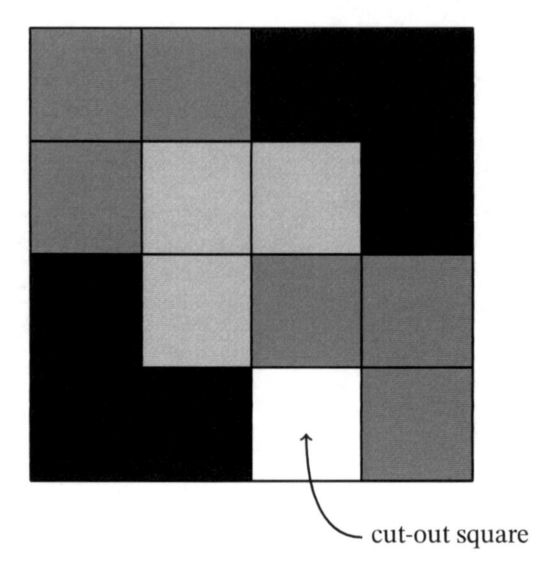

Figure 4.1. One possible covering for the case involving $n = 2$ for Problem 4.8.

4.2 More on Induction

In the previous section we discussed proving statements of the form $(\forall n \in \mathbb{N})P(n)$. Mathematical induction can actually be used to prove a broader family of results; namely, those of the form

$$(\forall n \in \mathbb{Z})(n \geq a \Longrightarrow P(n))$$

for any value $a \in \mathbb{Z}$. Theorem 4.2 handles the special case when $a = 1$. The ladder analogy from the previous section holds for this more general situation, too. To prove the next theorem, mimic the proof of Theorem 4.2, but this time use the set $S = \{k \in \mathbb{N} \mid P(a + k - 1) \text{ is true}\}$.

Theorem 4.9 (Principle of Mathematical Induction). *Let $P(a), P(a+1), P(a+2), \ldots$ be a sequence of statements, one for each integer greater than or equal to a. Assume that*

(i) *$P(a)$ is true, and*

(ii) *if $P(k)$ is true, then $P(k+1)$ is true.*

Then $P(n)$ is true for all integers $n \geq a$.

Theorem 4.9 gives a process for proving statements of the form: "For all integers $n \geq a$, $P(n)$." As before, hypothesis (i) is called the **base step**, and (ii) is called the **inductive step**.

Skeleton Proof 4.10 (Proof of $(\forall n \in \mathbb{Z})(n \geq a \implies P(n))$ by Induction)**.** Here is the general structure for a proof by induction when the base case does not necessarily involve $a = 1$.

Proof. We proceed by induction.

(i) Base step: [*Verify that $P(a)$ is true. This often, but not always, amounts to plugging $n = a$ into two sides of some claimed equation and verifying that both sides are actually equal.*]

(ii) Inductive step: [*Your goal is to prove "For all $k \in \mathbb{Z}$, if $P(k)$ is true, then $P(k+1)$ is true."*] Let $k \geq a$ be an integer and assume that $P(k)$ is true. [*Do something to derive that $P(k+1)$ is true.*] Therefore, $P(k+1)$ is true.

Thus, by induction, $P(n)$ is true for all integers $n \geq a$. □

We encountered the next theorem back in Section 3.3 (see Conjecture 3.29), but we did not prove it. When proving this theorem using induction, you will need to argue that if you add one more element to a finite set, then you end up with twice as many subsets. For your base case, consider the empty set.

Theorem 4.11. *If A is a finite set with n elements, then $\mathcal{P}(A)$ is a set with 2^n elements.*

Theorem 4.12. *For all integers $n \geq 0$, $n < 2^n$.*

One consequence of the previous two theorems is that the power set of a finite set always consists of more elements than the original set.

Theorem 4.13. *For all integers $n \geq 0$, 4 divides $9^n - 5$.*

Theorem 4.14. *For all integers $n \geq 0$, 4 divides $6 \cdot 7^n - 2 \cdot 3^n$.*

Theorem 4.15. *For all integers $n \geq 2$, $2^n > n + 1$.*

Theorem 4.16. *For all integers $n \geq 0$, $1 + 2^1 + 2^2 + \cdots + 2^n = 2^{n+1} - 1$.*

Theorem 4.17. *Fix a real number $r \neq 1$. For all integers $n \geq 0$,*

$$1 + r^1 + r^2 + \cdots + r^n = \frac{r^{n+1} - 1}{r - 1}.$$

Theorem 4.18. *For all integers $n \geq 3$,*

$$2 \cdot 3 + 3 \cdot 4 + \cdots + (n-1) \cdot n = \frac{(n-2)(n^2 + 2n + 3)}{3}.$$

Theorem 4.19. *For all integers* $n \geq 1$, $\dfrac{1}{1 \cdot 2} + \dfrac{1}{2 \cdot 3} + \cdots + \dfrac{1}{n(n+1)} = \dfrac{n}{n+1}$.

Theorem 4.20. *For all integers* $n \geq 1$,

$$\frac{1}{1 \cdot 3} + \frac{1}{3 \cdot 5} + \frac{1}{5 \cdot 7} + \cdots + \frac{1}{(2n-1)(2n+1)} = \frac{n}{2n+1}.$$

Theorem 4.21. *For all integers* $n \geq 0$, $3^{2n} - 1$ *is divisible by* 8.

Theorem 4.22. *For all integers* $n \geq 2$, $2^n < (n+1)!$.

Theorem 4.23. *For all integers* $n \geq 2$, $2 \cdot 9^n - 10 \cdot 3^n$ *is divisible by* 4.

We now consider an induction problem of a different sort, where you have to begin with some experimentation. For Part (c), consider using the results from Parts (a) and (b).

Problem 4.24. Suppose n lines are drawn in the plane so that no two lines are parallel and no three lines intersect at any one point. Such a collection of lines is said to be in **general position**. Every collection of lines in general position divides the plane into disjoint regions, some of which are polygons with finite area (bounded regions) and some of which are not (unbounded regions).

(a) Let $R(n)$ be the number of regions the plane is divided into by n lines in general position. Conjecture a formula for $R(n)$ and prove that your conjecture is correct.

(b) Let $U(n)$ be the number of unbounded regions the plane is divided into by n lines in general position. Conjecture a formula for $U(n)$ and prove that your conjecture is correct.

(c) Let $B(n)$ be the number of bounded regions the plane is divided into by n lines in general position. Conjecture a formula for $B(n)$ and prove that your conjecture is correct.

(d) Suppose we color each of the regions (bounded and unbounded) so that no two adjacent regions (i.e., share a common edge) have the same color. What is the fewest colors we could use to accomplish this? Prove your assertion.

> If you don't learn to fail, you will fail to learn.
>
> Manu Kapur, learning scientist

4.3 Complete Induction

There is another formulation of induction, where the inductive step begins with a set of assumptions rather than one single assumption. This method is sometimes called **complete induction** or **strong induction**.

Theorem 4.25 (Principle of Complete Mathematical Induction). *Let $P(1), P(2), P(3), \ldots$ be a sequence of statements, one for each natural number. Assume that*

(i) *$P(1)$ is true, and*

(ii) *For all $k \in \mathbb{N}$, if $P(j)$ is true for all $j \in \mathbb{N}$ such that $j \leq k$, then $P(k+1)$ is true.*

Then $P(n)$ is true for all $n \in \mathbb{N}$.

Note the difference between ordinary induction (Theorems 4.2 and 4.9) and complete induction. For the induction step of complete induction, we are not only assuming that $P(k)$ is true, but rather that $P(j)$ is true for all j from 1 to k. Despite the name, complete induction is not any stronger or more powerful than ordinary induction. It is worth pointing out that anytime ordinary induction is an appropriate proof technique, so is complete induction. So, when should we use complete induction?

In the inductive step, you need to reach $P(k+1)$, and you should ask yourself which of the previous cases you need to get there. If all you need, is the statement $P(k)$, then ordinary induction is the way to go. If two preceding cases, $P(k-1)$ and $P(k)$, are necessary to reach $P(k+1)$, then complete induction is appropriate. In the extreme, if one needs the full range of preceding cases (i.e., all statements $P(1), P(2), \ldots, P(k)$), then again complete induction should be utilized.

Note that in situations where complete induction is appropriate, it might be the case that you need to verify more than one case in the base step. The number of base cases to be checked depends on how one needs to "look back" in the induction step.

Skeleton Proof 4.26 (Proof of $(\forall n \in \mathbb{N})P(n)$ by Complete Induction). Here is the general structure for a proof by complete induction.

Proof. We proceed by induction.

(i) Base step: [*Verify that $P(1)$ is true. Depending on the statement, you may also need to verify that $P(k)$ is true for other specific values of k.*]

(ii) Inductive step: [*Your goal is to prove "For all $k \in \mathbb{N}$, if for each $k \in \mathbb{N}$, $P(j)$ is true for all $j \in \mathbb{N}$ such that $j \leq k$, then $P(k+1)$ is true."*] Let $k \in \mathbb{N}$. Suppose $P(j)$ is true for all $j \leq k$. [*Do something to derive that $P(k+1)$ is true.*] Therefore, $P(k+1)$ is true.

Thus, by complete induction, $P(n)$ is true for all integers $n \geq a$. □

When tackling the problems in this section, think carefully about how many base steps you must verify.

Theorem 4.27. *Define a sequence of numbers by* $a_1 = 1$, $a_2 = 3$, *and* $a_n = 3a_{n-1} - 2a_{n-2}$ *for all natural numbers* $n \geq 3$. *Then* $a_n = 2^n - 1$ *for all* $n \in \mathbb{N}$.

Theorem 4.28. *Define a sequence of numbers by* $a_1 = 3, a_2 = 5, a_3 = 9$, *and* $a_n = 2a_{n-1} + a_{n-2} - 2a_{n-3}$ *for all natural numbers* $n \geq 4$. *Then* $a_n = 2^n + 1$ *for all* $n \in \mathbb{N}$.

Problem 4.29. The Fibonacci sequence is given by $f_1 = 1$, $f_2 = 1$, and $f_n = f_{n-1} + f_{n-2}$ for all natural numbers $n \geq 3$. Prove that $\left(\frac{3}{2}\right)^{n-2} \leq f_n \leq 2^n$ for all $n \in \mathbb{N}$.

Recall that Theorem 4.9 generalized Theorem 4.2 and allowed us to handle situations where the base case was something other than $P(1)$. We can generalize complete induction in the same way, but we will not write this down as a formal theorem.

Problem 4.30. Prove that every amount of postage that is at least 12 cents can be made from 4-cent and 5-cent stamps.

Problem 4.31. Whoziwhatzits come in boxes of 6, 9, and 20. Prove that for any natural number $n \geq 44$, it is possible to buy exactly n Whoziwhatzits with a combination of these boxes.

Problem 4.32. Consider a grid of squares that is 2 squares wide and n squares long. Using n dominoes that are 1 square by 2 squares, there are many ways to perfectly cover this chessboard with no overlap. How many? Prove your answer.

Problem 4.33. A binary string of length n is an ordered list of n digits such that each digit is either 0 or 1. For example 011101 and 011011 are distinct binary strings of length 6. Here are the rules for *Binary Solitaire*: At any stage, you are allowed to:

(i) Swap the leftmost digit (i.e., change 0 to 1, or 1 to 0). For example, we can do 011101 → 11101.

(ii) Swap the the digit immediately to the right of the leftmost occurrence of 1. For example, we can do 011011 → 010011.

Prove that for all $n \in \mathbb{N}$, you can change any binary string of length n to any other binary string of the same length.

Problem 4.34. Prove that the number of binary strings of length n that never have two consecutive 1's is the Fibonacci number f_{n+2}. See Problem 4.29 for the definition of the Fibonacci numbers.

> Nothing that's worth anything is ever easy.
>
> Mike Hall, ultra-distance cyclist

4.4 The Well-Ordering Principle

The penultimate theorem of this chapter is known as the **Well-Ordering Principle**. As you shall see, this seemingly obvious theorem requires a bit of work to prove. It is worth noting that in some axiomatic systems, the Well-Ordering Principle is sometimes taken as an axiom. However, in our case, the result follows from complete induction. Before stating the Well-Ordering Principle, we need an additional definition.

Definition 4.35. Let $A \subseteq \mathbb{R}$ and $m \in A$. Then m is called a **maximum** (or **greatest element**) of A if for all $a \in A$, we have $a \leq m$. Similarly, m is called **minimum** (or **least element**) of A if for all $a \in A$, we have $m \leq a$.

Not surprisingly, maximums and minimums are unique when they exist. It might be helpful to review Skeleton Proof 2.90 prior to attacking the next result.

Theorem 4.36. *If $A \subseteq \mathbb{R}$ such that the maximum (respectively, minimum) of A exists, then the maximum (respectively, minimum) of A is unique.*

If the maximum of a set A exists, then it is denoted by $\boxed{\max(A)}$. Similarly, if the minimum of a set A exists, then it is denoted by $\boxed{\min(A)}$.

Problem 4.37. Find the maximum and the minimum for each of the following sets when they exist.

(a) $\{5, 11, 17, 42, 103\}$

(b) $\mathbb{N}$

(c) $\mathbb{Z}$

(d) $(0, 1]$

(e) $(0, 1] \cap \mathbb{Q}$

(f) $(0, \infty)$

(g) $\{42\}$

(h) $\{\frac{1}{n} \mid n \in \mathbb{N}\}$

(i) $\{\frac{1}{n} \mid n \in \mathbb{N}\} \cup \{0\}$

(j) $\emptyset$

To prove the Well-Ordering Principle, consider a proof by contradiction. Suppose S is a nonempty subset of $\mathbb{N}$ that does not have a least element. Define the proposition $P(n)$:="n is not an element of S" and then use complete induction to prove the result.

Theorem 4.38 (Well-Ordering Principle). *Every nonempty subset of the natural numbers has a least element.*

It turns out that the Well-Ordering Principle (Theorem 4.38) and the Axiom of Induction (Axiom 4.1) are equivalent. In other words, one can prove the Well-Ordering Principle from the Axiom of Induction, as we have done, but one can also prove the Axiom of Induction if the Well-Ordering Principle is assumed.

The final two theorems of this section can be thought of as generalized versions of the Well-Ordering Principle.

Theorem 4.39. *If A is a nonempty subset of the integers and there exists $l \in \mathbb{Z}$ such that $l \leq a$ for all $a \in A$, then A contains a least element.*

Theorem 4.40. *If A is a nonempty subset of the integers and there exists $u \in \mathbb{Z}$ such that $a \leq u$ for all $a \in A$, then A contains a greatest element.*

The element l in Theorem 4.39 is referred to as a **lower bound** for A while the element u in Theorem 4.40 is called an **upper bound** for A. We will study lower and upper bounds in more detail in Section 5.1.

> Life is like riding a bicycle. To keep your balance you must keep moving.
>
> Albert Einstein, theoretical physicist

> All truths are easy to understand once they are discovered; the point is to discover them.
>
> Galileo Galilei, astronomer & physicist

5

The Real Numbers

In this chapter we will take a deep dive into structure of the real numbers by building up the multitude of properties you are familiar with by starting with a collection of fundamental axioms. Recall that an axiom is a statement that is assumed to be true without proof. These are the basic building blocks from which all theorems are proved. It is worth pointing out that one can carefully construct the real numbers from the natural numbers. However, that will not be the approach we take. Instead, we will simply list the axioms that the real numbers satisfy.

5.1 Axioms of the Real Numbers

Our axioms for the real numbers fall into three categories:

(1) **Field Axioms:** These axioms provide the essential properties of arithmetic involving addition and subtraction.

(2) **Order Axioms:** These axioms provide the necessary properties of inequalities.

(3) **Completeness Axiom:** This axiom ensures that the familiar number line that we use to model the real numbers does not have any holes in it.

We begin with the Field Axioms.

Axioms 5.1 (Field Axioms). *There exist operations* $+$ *(addition) and* $\cdot$ *(multiplication) on* $\mathbb{R}$ *satisfying:*

(F1) *(Associativity for Addition) For all* $a, b, c \in \mathbb{R}$ *we have* $(a+b)+c = a+(b+c)$*;*

(F2) *(Commutativity for Addition) For all* $a, b \in \mathbb{R}$*, we have* $a + b = b + a$*;*

(F3) *(Additive Identity) There exists* $0 \in \mathbb{R}$ *such that for all* $a \in \mathbb{R}$*,* $0 + a = a$*;*

(F4) *(Additive Inverses) For all* $a \in \mathbb{R}$ *there exists* $-a \in \mathbb{R}$ *such that* $a + (-a) = 0$*;*

(F5) *(Associativity for Multiplication) For all* $a, b, c \in \mathbb{R}$ *we have* $(ab)c = a(bc)$*;*

(F6) *(Commutativity for Multiplication) For all* $a, b \in \mathbb{R}$*, we have* $ab = ba$*;*

(F7) *(Multiplicative Identity) There exists* $1 \in \mathbb{R}$ *such that* $1 \neq 0$ *and for all* $a \in \mathbb{R}$*,* $1a = a$*;*

(F8) *(Multiplicative Inverses) For all* $a \in \mathbb{R} \setminus \{0\}$ *there exists* $a^{-1} \in \mathbb{R}$ *such that* $aa^{-1} = 1$*.*

(F9) *(Distributive Property) For all* $a, b, c \in \mathbb{R}$*,* $a(b + c) = ab + ac$*;*

In the language of abstract algebra, Axioms F1–F4 and F5–F8 make each of $\mathbb{R}$ and $\mathbb{R} \setminus \{0\}$ an abelian group under addition and multiplication, respectively. Axiom F9 provides a way for the operations of addition and multiplication to interact. Collectively, Axioms F1–F9 make the real numbers a **field**. It follows from the axioms that the elements 0 and 1 of $\mathbb{R}$ are the unique **additive** and **multiplicative identities** in $\mathbb{R}$. To prove the following theorem, suppose 0 and $0'$ are both additive identities in $\mathbb{R}$ and then show that $0 = 0'$. This shows that there can only be one additive identity.

Theorem 5.2. *The additive identity of* $\mathbb{R}$ *is unique.*

To prove the next theorem, mimic the approach you used to prove Theorem 5.2.

Theorem 5.3. *The multiplicative identity of* $\mathbb{R}$ *is unique.*

For every $a \in \mathbb{R}$, the elements $-a$ and a^{-1} (as long as $a \neq 0$) are also the unique **additive** and **multiplicative inverses**, respectively.

Theorem 5.4. *Every real number has a unique additive inverse.*

Theorem 5.5. *Every nonzero real number has a unique multiplicative inverse.*

Since we are taking a formal axiomatic approach to the real numbers, we should make it clear how the natural numbers are embedded in $\mathbb{R}$.

Definition 5.6. We define the **natural numbers**, denoted by $\mathbb{N}$, to be the smallest subset of $\mathbb{R}$ satisfying:

(a) $1 \in \mathbb{N}$, and

(b) for all $n \in \mathbb{N}$, we have $n + 1 \in \mathbb{N}$.

Notice the similarity between the definition of the natural numbers presented above and the Axiom of Induction given in Section 4.1. Of course, we use the standard numeral system to represent the natural numbers, so that $\mathbb{N} = \{1, 2, 3, 4, 5, 6, 7, 8, 9, 10 \ldots\}$.

Given the natural numbers, Axiom F3/Theorem 5.2 and Axiom F4/Theorem 5.4 together with the operation of addition allow us to define the **integers**, denoted by $\mathbb{Z}$, in the obvious way. That is, the integers consist of the natural numbers together with the additive identity and all of the additive inverses of the natural numbers.

We now introduce some common notation that you are likely familiar with. Take a moment to think about why the following is a definition as opposed to an axiom or theorem.

Definition 5.7. For every $a, b \in \mathbb{R}$ and $n \in \mathbb{Z}$, we define the following:

(a) $\boxed{a - b := a + (-b)}$

(b) $\boxed{\dfrac{a}{b} := ab^{-1}}$ (for $b \neq 0$)

(c) $$\boxed{a^n := \begin{cases} \overbrace{aa \cdots a}^{n}, & \text{if } n \in \mathbb{N} \\ 1, & \text{if } n = 0 \text{ and } a \neq 0 \\ \dfrac{1}{a^{-n}}, & \text{if } -n \in \mathbb{N} \text{ and } a \neq 0 \end{cases}}$$

The set of **rational numbers**, denoted by $\mathbb{Q}$, is defined to be the collection of all real numbers having the form given in Part (b) of Definition 5.7. The **irrational numbers** are defined to be $\mathbb{R} \setminus \mathbb{Q}$.

Using the Field Axioms, we can prove each of the statements in the following theorem.

Theorem 5.8. *For all $a, b, c \in \mathbb{R}$, we have the following:*

(a) *$a = b$ if and only if $a + c = b + c$;*

(b) *$0a = 0$;*

(c) *$-a = (-1)a$;*

(d) $(-1)^2 = 1$;

(e) $-(-a) = a$;

(f) *If* $a \neq 0$, *then* $(a^{-1})^{-1} = a$;

(g) *If* $a \neq 0$ *and* $ab = ac$, *then* $b = c$.

(h) *If* $ab = 0$, *then either* $a = 0$ *or* $b = 0$.

Carefully prove the next theorem by explicitly citing where you are utilizing the Field Axioms and Theorem 5.8.

Theorem 5.9. *For all* $a, b \in \mathbb{R}$, *we have* $(a + b)(a - b) = a^2 - b^2$.

We now introduce the Order Axioms of the real numbers.

Axioms 5.10 (Order Axioms). *For* $a, b, c \in \mathbb{R}$, *there is a relation* $\boxed{<}$ *on* $\mathbb{R}$ *satisfying:*

(O1) (*Trichotomy Law*) *If* $a \neq b$, *then either* $a < b$ *or* $b < a$ *but not both;*

(O2) (*Transitivity*) *If* $a < b$ *and* $b < c$, *then* $a < c$;

(O3) *If* $a < b$, *then* $a + c < b + c$;

(O4) *If* $a < b$ *and* $0 < c$, *then* $ac < bc$;

Given Axioms O1–O4, we say that the real numbers are a **linearly ordered field**. We call numbers greater than zero **positive** and those greater than or equal to zero **nonnegative**. There are similar definitions for **negative** and **nonpositive**.

Notice that the Order Axioms are phrased in terms of "$<$". We would also like to be able to utilize "$>$", "$\leq$", and "$\geq$".

Definition 5.11. For $a, b \in \mathbb{R}$, we define:

(a) $\boxed{a > b}$ if $b < a$;

(b) $\boxed{a \leq b}$ if $a < b$ or $a = b$;

(c) $\boxed{a \geq b}$ if $b \leq a$.

Notice that we took the existence of the inequalities "$<$", "$>$", "$\leq$", and "$\geq$" on the real numbers for granted when we defined intervals of real numbers in Definition 3.4.

Using the Order Axioms, we can prove many familiar facts.

Theorem 5.12. *For all $a, b \in \mathbb{R}$, if $a, b > 0$, then $a + b > 0$; and if $a, b < 0$, then $a + b < 0$.*

The next result extends Axiom O3.

Theorem 5.13. *For all $a, b, c, d \in \mathbb{R}$, if $a < b$ and $c < d$, then $a + c < b + d$.*

Theorem 5.14. *For all $a \in \mathbb{R}$, $a > 0$ if and only if $-a < 0$.*

Theorem 5.15. *If a, b, c, and d are positive real numbers such that $a < b$ and $c < d$, then $ac < bd$.*

Theorem 5.16. *For all $a, b \in \mathbb{R}$, we have the following:*

(a) *$ab > 0$ if and only if either $a, b > 0$ or $a, b < 0$;*

(b) *$ab < 0$ if and only if $a < 0 < b$ or $b < 0 < a$.*

Theorem 5.17. *For all positive real numbers a and b, $a < b$ if and only if $a^2 < b^2$.*

Consider using three cases when approaching the proof of the following theorem.

Theorem 5.18. *For all $a \in \mathbb{R}$, we have $a^2 \geq 0$.*

It might come as a surprise that the following result requires proof.

Theorem 5.19. *We have $0 < 1$.*

The previous theorem together with Theorem 5.14 implies that $-1 < 0$ as you expect. It also follows from Axiom O3 that for all $n \in \mathbb{Z}$, we have $n < n + 1$. We assume that there are no integers between n and $n + 1$.

Theorem 5.20. *For all $a \in \mathbb{R}$, if $a > 0$, then $a^{-1} > 0$, and if $a < 0$, then $a^{-1} < 0$.*

Theorem 5.21. *For all $a, b \in \mathbb{R}$, if $a < b$, then $-b < -a$.*

The last few results allow us to take for granted our usual understanding of which real numbers are positive and which are negative. The next theorem yields a result that extends Theorem 5.21.

Theorem 5.22. *For all $a, b, c \in \mathbb{R}$, if $a < b$ and $c < 0$, then $bc < ac$.*

There is a special function that we can now introduce.

Definition 5.23. Given $a \in \mathbb{R}$, we define the **absolute value of** a, denoted $|a|$, via

$$|a| := \begin{cases} a, & \text{if } a \geq 0 \\ -a, & \text{if } a < 0. \end{cases}$$

Theorem 5.24. *For all $a \in \mathbb{R}$, $|a| \geq 0$ with equality only if $a = 0$.*

We can interpret $|a|$ as the distance between a and 0 as depicted in Figure 5.1.

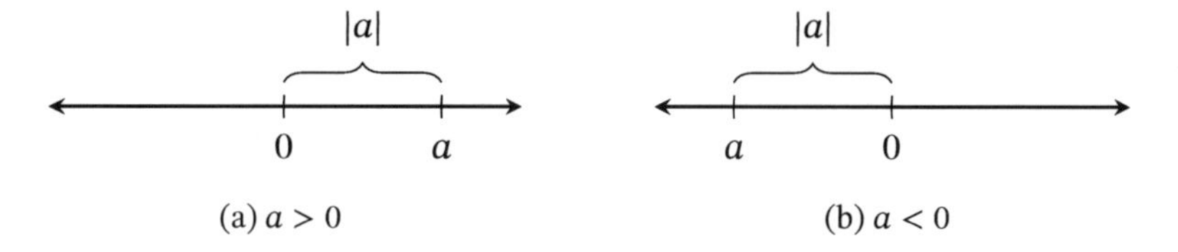

Figure 5.1. Visual representation of $|a|$.

Theorem 5.25. *For all $a, b \in \mathbb{R}$, we have $|a - b| = |b - a|$.*

Given two points a and b, $|a - b|$, and hence $|b - a|$ by the previous theorem, is the distance between a and b as shown in Figure 5.2.

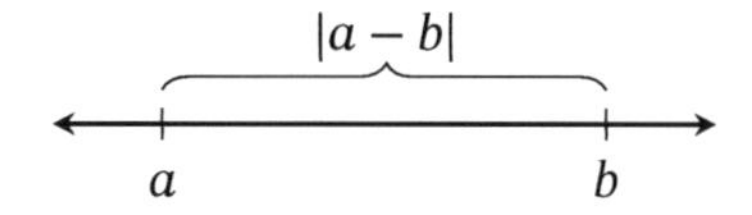

Figure 5.2. Visual representation of $|a - b|$.

Theorem 5.26. *For all $a, b \in \mathbb{R}$, $|ab| = |a||b|$.*

In the next theorem, writing $\pm a \leq b$ is an abbreviation for $a \leq b$ and $-a \leq b$.

Theorem 5.27. *For all $a, b \in \mathbb{R}$, if $\pm a \leq b$, then $|a| \leq b$.*

Theorem 5.28. *For all $a \in \mathbb{R}$, $|a|^2 = a^2$.*

Theorem 5.29. *For all $a \in \mathbb{R}$, $\pm a \leq |a|$.*

Theorem 5.30. *For all $a, r \in \mathbb{R}$ with r nonnegative, $|a| \leq r$ if and only if $-r \leq a \leq r$.*

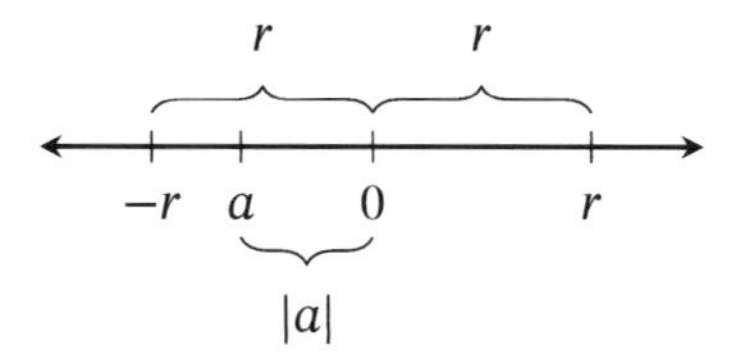

Figure 5.3. Visual representation of $|a| \leq r$.

The letter r was used in the previous theorem because it is the first letter of the word "radius". If r is positive, we can think of the interval $(-r, r)$ as the interior of a one-dimensional circle with radius r centered at 0. Figure 5.3 provides a visual interpretation of Theorem 5.30.

Corollary 5.31. *For all $a, b, r \in \mathbb{R}$ with r nonnegative, $|a - b| \leq r$ if and only if $b - r \leq a \leq b + r$.*

Since $|a - b|$ represents the distance between a and b, we can interpret $|a - b| \leq r$ as saying that the distance between a and b is less than or equal to r. In other words, a is within r units of b. See Figure 5.4.

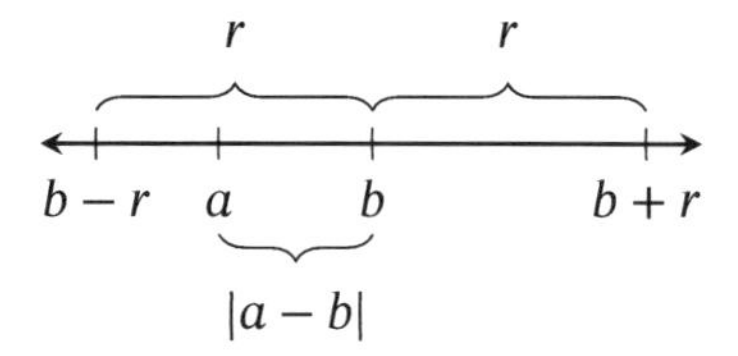

Figure 5.4. Visual representation of $|a - b| \leq r$.

Consider using Theorems 5.29 and 5.30 when attacking the next result, which is known as the **Triangle Inequality**. This result can be extremely useful in some contexts.

Theorem 5.32 (Triangle Inequality). *For all $a, b \in \mathbb{R}$, $|a + b| \leq |a| + |b|$.*

Figure 5.5 depicts two of the cases for the Triangle Inequality.

Problem 5.33. Under what conditions do we have equality for the Triangle Inequality?

Where did the Triangle Inequality get its name? Why "Triangle"? For any triangle (including degenerate triangles), the sum of the lengths of any two sides

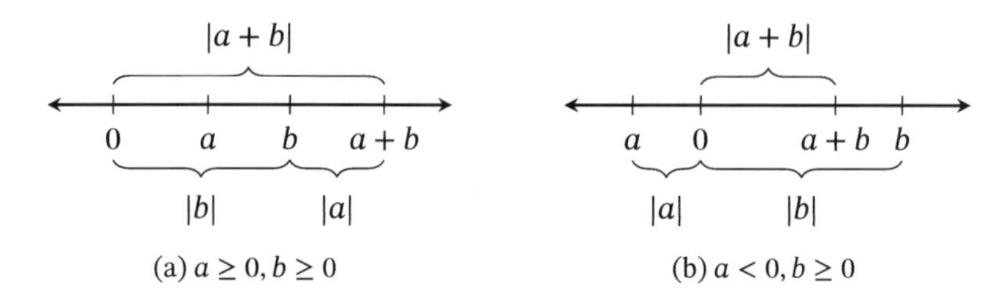

Figure 5.5. Visual representation of two of the cases for the Triangle Inequality.

must be greater than or equal to the length of the remaining side. That is, if x, y, and z are the lengths of the sides of the triangle, then $z \leq x + y$, where we have equality only in the degenerate case of a triangle with no area. In linear algebra, the Triangle Inequality is a theorem about lengths of vectors. If $\mathbf{a}$ and $\mathbf{b}$ are vectors in $\mathbb{R}^n$, then the Triangle Inequality states that $||\mathbf{a} + \mathbf{b}|| \leq ||\mathbf{a}|| + ||\mathbf{b}||$. Note that $||\mathbf{a}||$ denotes the length of vector $\mathbf{a}$. See Figure 5.6. The version of the Triangle Inequality that we presented in Theorem 5.32 is precisely the one-dimensional version of the Triangle Inequality in terms of vectors.

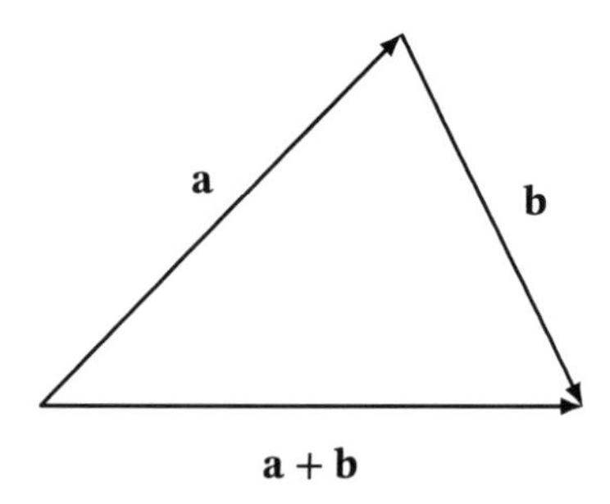

Figure 5.6. Triangle Inequality in terms of vectors.

The next theorem is sometimes called the **Reverse Triangle Inequality**.

Theorem 5.34 (Reverse Triangle Inequality). *For all* $a, b \in \mathbb{R}$, $|a-b| \geq ||a| - |b||$.

Before we introduce the Completeness Axiom, we need some additional terminology.

Definition 5.35. Let $A \subseteq \mathbb{R}$. A point b is called an **upper bound** of A if for all $a \in A$, $a \leq b$. The set A is said to be **bounded above** if it has an upper bound.

Problem 5.36. The notion of a **lower bound** and the property of a set being **bounded below** are defined similarly. Try defining them.

Problem 5.37. Find all upper bounds and all lower bounds for each of the following sets when they exist.

(a) $\{5, 11, 17, 42, 103\}$

(b) $\mathbb{N}$

(c) $\mathbb{Z}$

(d) $(0, 1]$

(e) $(0, 1] \cap \mathbb{Q}$

(f) $(0, \infty)$

(g) $\{42\}$

(h) $\{\frac{1}{n} \mid n \in \mathbb{N}\}$

(i) $\{\frac{1}{n} \mid n \in \mathbb{N}\} \cup \{0\}$

(j) $\emptyset$

Definition 5.38. A set $A \subseteq \mathbb{R}$ is **bounded** if A is bounded above and below.

Notice that a set $A \subseteq \mathbb{R}$ is bounded if and only if it is a subset of some bounded closed interval.

Definition 5.39. Let $A \subseteq \mathbb{R}$. A point p is a **supremum** (or **least upper bound**) of A if p is an upper bound of A and $p \leq b$ for every upper bound b of A. Analogously, a point p is an **infimum** (or **greatest lower bound**) of A if p is a lower bound of A and $p \geq b$ for every lower bound b of A.

Our next result tells us that a supremum of a set and an infimum of a set are unique when they exist.

Theorem 5.40. *If $A \subseteq \mathbb{R}$ such that a supremum (respectively, infimum) of A exists, then the supremum (respectively, infimum) of A is unique.*

In light of the previous theorem, if the supremum of A exists, it is denoted by $\boxed{\sup(A)}$. Similarly, if the infimum of A exists, it is denoted by $\boxed{\inf(A)}$.

Problem 5.41. Find the supremum and the infimum of each of the sets in Problem 5.37 when they exist.

It is important to recognize that the supremum or infimum of a set may or may not be contained in the set. In particular, we have the following theorem concerning suprema and maximums. The analogous result holds for infima and minimums.

Theorem 5.42. *Let $A \subseteq \mathbb{R}$. Then A has a maximum if and only if A has a supremum and* $\sup(A) \in A$, *in which case the* $\max(A) = \sup(A)$.

Intuitively, a point is the supremum of a set A if and only if no point smaller than the supremum can be an upper bound of A. The next result makes this more precise.

Theorem 5.43. *Let $A \subseteq \mathbb{R}$ such that A is bounded above and let b be an upper bound of A. Then b is the supremum of A if and only if for every $\varepsilon > 0$, there exists $a \in A$ such that $b - \varepsilon < a$.*

Problem 5.44. State and prove the analogous result to Theorem 5.43 involving infimum.

The following axiom states that every nonempty subset of the real numbers that has an upper bound has a least upper bound.

Axiom 5.45 (Completeness Axiom). *If A is a nonempty subset of $\mathbb{R}$ that is bounded above, then* $\sup(A)$ *exists.*

Given the Completeness Axiom, we say that the real numbers satisfy the **least upper bound property**. It is worth mentioning that we do not need the Completeness Axiom to conclude that every nonempty subset of the integers that is bounded above has a supremum, as this follows from Theorem 4.40 (a generalized version of the Well-Ordering Principle).

Certainly, the real numbers also satisfy the analogous result involving infimum.

Theorem 5.46. *If A is a nonempty subset of $\mathbb{R}$ that is bounded below, then* $\inf(A)$ *exists.*

Our next result, called the **Archimedean Property**, tells us that for every real number, we can always find a natural number that is larger. To prove this theorem, consider a proof by contradiction and then utilize the Completeness Axiom and Theorem 5.43.

Theorem 5.47 (Archimedean Property). *For every $x \in \mathbb{R}$, there exists $n \in \mathbb{N}$ such that $x < n$.*

More generally, we can "squeeze" every real number between a pair of integers. The next result is sometimes referred to at the **Generalized Archimedean Property**.

Theorem 5.48 (Generalized Archimedean Property). *For every $x \in \mathbb{R}$, there exists $k, n \in \mathbb{Z}$ such that $k < x < n$.*

Theorem 5.49. *For any positive real number x, there exists $N \in \mathbb{N}$ such that $0 < \frac{1}{N} < x$.*

The next theorem strengthens the Generalized Archimedean Property and says that every real number is either an integer or lies between a pair of consecutive integers. To prove this theorem, let $x \in \mathbb{R}$ and define $L = \{k \in \mathbb{Z} \mid k \leq x\}$. Use the Generalized Archimedean Property to conclude that L is nonempty and then utilize Theorem 4.40.

Theorem 5.50. *For every $x \in \mathbb{R}$, there exists $n \in \mathbb{Z}$ such that $n \leq x < n + 1$.*

To prove the next theorem, let $a < b$, utilize Theorem 5.49 on $b - a$ to obtain $N \in \mathbb{N}$ such that $\frac{1}{N} < b - a$, and then apply Theorem 5.50 to Na to conclude that there exists $n \in \mathbb{N}$ such that $n \leq Na < n + 1$. Lastly, argue that $\frac{n+1}{N}$ is the rational number you seek.

Theorem 5.51. *If (a, b) is an open interval, then there exists a rational number p such that $p \in (a, b)$.*

Recall that the real numbers consist of rational and irrational numbers. Two examples of an irrational number that you are likely familiar with are π and $\sqrt{2}$. In Section 6.2, we will prove that $\sqrt{2}$ is irrational, but for now we will take this fact for granted. It turns out that $\sqrt{2} \approx 1.41421356237 \in (1, 2)$. This provides an example of an irrational number occurring between a pair of distinct rational numbers. The following theorem is a good challenge to generalize this.

Theorem 5.52. *If (a, b) is an open interval, then there exists an irrational number p such that $p \in (a, b)$.*

Repeated applications of the previous two theorems implies that every open interval contains infinitely many rational numbers and infinitely many irrational numbers. In light of these two theorems, we say that both the rationals and irrationals are **dense** in the real numbers.

> If people do not believe that mathematics is simple, it is only because they do not realize how complicated life is.
>
> John von Neumann, mathematician

5.2 Standard Topology of the Real Line

In this section, we will introduce the notions of open, closed, compact, and connected as they pertain to subsets of the real numbers. These properties form the underpinnings of a branch of mathematics called **topology** (derived from the Greek words *tópos*, meaning 'place, location', and *ology*, meaning 'study of'). Topology, sometimes called "rubber sheet geometry," is concerned with properties of spaces that are invariant under any continuous deformation (e.g., bending, twisting, and stretching like rubber while not allowing tearing apart or gluing together). The fundamental concepts in topology are continuity, compactness, and connectedness, which rely on ideas such as "arbitrary close" and "far apart". These ideas can be made precise using open sets.

Once considered an abstract branch of pure mathematics, topology now has applications in biology, computer science, physics, and robotics. The goal of this section is to introduce you to the basics of the set-theoretic definitions used in topology and to provide you with an opportunity to tinker with open and closed subsets of the real numbers. In Section 8.5, we will revisit these concepts and explore continuous functions.

For this entire section, our universe of discourse is the set of real numbers. You may assume all the usual basic algebraic properties of the real numbers (addition, subtraction, multiplication, division, commutative property, distribution, etc.). We will often refer to an element in a subset of real numbers as a **point**.

Definition 5.53. A set U is called an **open set** if for every $x \in U$, there exists a bounded open interval (a, b) containing x such that $(a, b) \subseteq U$.

It follows immediately from the definition that every open set is a union of bounded open intervals.

Problem 5.54. Determine whether each of the following sets is open. Justify your assertions.

(a) $(1, 2)$

(b) $(1, \infty)$

(c) $(1, 2) \cup (\pi, 5)$

(d) $[1, 2]$

(e) $(-\infty, \sqrt{2}]$

(f) $\{4, 17, 42\}$

(g) $\{\frac{1}{n} \mid n \in \mathbb{N}\}$

(h) $\{\frac{1}{n} \mid n \in \mathbb{N}\} \cup \{0\}$

(i) $\mathbb{R}$

(j) $\mathbb{Q}$

(k) $\mathbb{Z}$

(l) $\emptyset$

As expected, every open interval (i.e., intervals of the form (a, b), $(-\infty, b)$, (a, ∞), or $(-\infty, \infty)$) is an open set.

Theorem 5.55. *Every open interval is an open set.*

However, it is important to point out that open sets can be more complicated than a single open interval.

Problem 5.56. Provide an example of an open set that is not a single open interval.

Theorem 5.57. *If U and V are open sets, then*

(a) *$U \cup V$ is an open set, and*

(b) *$U \cap V$ is an open set.*

According to the next two theorems, the union of arbitrarily many open sets is open while the intersection of a finite number of open sets is open.

Theorem 5.58. *If $\{U_\alpha\}_{\alpha\in\Delta}$ is a collection of open sets, then $\bigcup_{\alpha\in\Delta} U_\alpha$ is an open set.*

Consider using induction to prove the next theorem.

Theorem 5.59. *If $\{U_i\}_{i=1}^n$ is a finite collection of open sets for $n \in \mathbb{N}$, then $\bigcap_{i=1}^n U_i$ is an open set.*

Problem 5.60. Explain why we cannot utilize induction to prove that the intersection of infinitely many open sets indexed by the natural numbers is open.

Problem 5.61. Give an example of each of the following.

(a) A collection of open sets $\{U_\alpha\}_{\alpha\in\Delta}$ such that $\bigcap_{\alpha\in\Delta} U_\alpha$ is an open set.

(b) A collection of open sets $\{U_\alpha\}_{\alpha\in\Delta}$ such that $\bigcap_{\alpha\in\Delta} U_\alpha$ is not an open set.

According to the previous problem, the intersection of infinitely many open sets may or may not be open. So, we know that there is no theorem that states that the intersection of arbitrarily many open sets is open. We only know for certain that the intersection of finitely many open sets is open by Theorem 5.59.

Definition 5.62. Suppose $A \subseteq \mathbb{R}$. A point $p \in \mathbb{R}$ is an **accumulation point** of A if for every bounded open interval (a, b) containing p, there exists a point $q \in (a, b) \cap A$ such that $q \neq p$.

Notice that if p is an accumulation point of A, then p may or may not be in A. Loosely speaking, p is an accumulation point of a set A if there are points in A arbitrarily close to p. That is, if we zoom in on p, we should always see points in A nearby.

Problem 5.63. Consider the open interval $I = (1, 2)$. Prove each of the following.

(a) The points 1 and 2 are accumulation points of I.

(b) If $p \in I$, then p is an accumulation point of I.

(c) If $p < 1$ or $p > 2$, then p is not an accumulation point of I.

Theorem 5.64. *A point p is an accumulation point of the intervals (a, b), $(a, b]$, $[a, b)$, and $[a, b]$ if and only if $p \in [a, b]$.*

Problem 5.65. Prove that the point $p = 0$ is an accumulation point of $A = \{\frac{1}{n} \mid n \in \mathbb{N}\}$. Are there any other accumulation points of A?

Problem 5.66. Provide an example of a set A with exactly two accumulation points.

Consider using Theorems 5.51 and 5.52 when proving the next result.

Theorem 5.67. *If $p \in \mathbb{R}$, then p is an accumulation point of $\mathbb{Q}$.*

Definition 5.68. A set $A \subseteq \mathbb{R}$ is called **closed** if A contains all of its accumulation points.

Problem 5.69. Determine whether each of the sets in Problem 5.54 is closed. Justify your assertions.

The upshot of Parts (i) and (l) of Problems 5.54 and 5.69 is that $\mathbb{R}$ and $\emptyset$ are both open and closed. It turns out that these are the only two subsets of the real numbers with this property. One issue with the terminology that could potentially create confusion is that the open interval $(-\infty, \infty)$ is both an open and a closed set.

Problem 5.70. Provide an example of each of the following. You do not need to prove that your answers are correct.

(a) A set that is open but not closed.

(b) A set that is closed but not open.

(c) A set that neither open nor closed.

Another potentially annoying feature of the terminology illustrated by Problem 5.70 is that if a set is not open, it may or may not be closed. Similarly, if a set is not closed, it may or may not be open. That is, open and closed are not opposites of each other.

The next result justifies referring to $[a, b]$ as a closed interval.

Theorem 5.71. *Every interval of the form* $[a, b]$, $(-\infty, b]$, $[a, \infty)$, *or* $(-\infty, \infty)$ *is a closed set.*

Theorem 5.72. *Every finite subset of* $\mathbb{R}$ *is closed.*

Despite the fact that open and closed are not opposites of each other, there is a nice relationship between open and closed sets in terms of complements.

Theorem 5.73. *Let* $U \subseteq \mathbb{R}$. *Then* U *is open if and only if* U^C *is closed.*

Theorem 5.74. *If* A *and* B *are closed sets, then*

(a) $A \cup B$ *is a closed set, and*

(b) $A \cap B$ *is a closed set.*

The next two theorems are analogous to Theorems 5.58 and 5.59.

Theorem 5.75. *If* $\{A_\alpha\}_{\alpha\in\Delta}$ *is a collection of closed sets, then* $\bigcap_{\alpha\in\Delta} A_\alpha$ *is a closed set.*

Theorem 5.76. *If* $\{A_i\}_{i=1}^n$ *is a finite collection of closed sets for* $n \in \mathbb{N}$, *then* $\bigcup_{i=1}^n A_i$ *is a closed set.*

Problem 5.77. Provide an example of a collection of closed sets $\{A_\alpha\}_{\alpha\in\Delta}$ such that $\bigcup_{\alpha\in\Delta} A_\alpha$ is not a closed set.

Problem 5.78. Determine whether each of the following sets is open, closed, both, or neither.

(a) $V = \displaystyle\bigcup_{n=2}^{\infty}\left(n - \frac{1}{2}, n\right)$

(b) $W = \displaystyle\bigcap_{n=2}^{\infty}\left(n - \frac{1}{2}, n\right)$

(c) $X = \displaystyle\bigcap_{n=1}^{\infty}\left(-\frac{1}{n}, \frac{1}{n}\right)$

(d) $Y = \bigcap_{n=1}^{\infty} (-n, n)$

(e) $Z = (0, 1) \cap \mathbb{Q}$

Problem 5.79. Prove or provide a counterexample: Every non-closed set has at least one accumulation point.

We now introduce three special classes of subsets of $\mathbb{R}$: compact, connected, and disconnected.

Definition 5.80. A set $K \subseteq \mathbb{R}$ is called **compact** if K is both closed and bounded.

It is important to point out that there is a more general definition of compact in an arbitrary topological space. However, using our notions of open and closed, it is a theorem that a subset of the real line is compact if and only if it is closed and bounded.

Problem 5.81. Determine whether each of the following sets is compact. Briefly justify your assertions.

(a) $[0, 1) \cup [2, 3]$

(b) $[0, 1) \cup (1, 2]$

(c) $[0, 1) \cup [1, 2]$

(d) $\mathbb{R}$

(e) $\mathbb{Q}$

(f) $\mathbb{R} \setminus \mathbb{Q}$

(g) $\mathbb{Z}$

(h) $\{\frac{1}{n} \mid n \in \mathbb{N}\}$

(i) $[0, 1] \cup \{1 + \frac{1}{n} \mid n \in \mathbb{N}\}$

(j) $\{17, 42\}$

(k) $\{17\}$

(l) $\emptyset$

Problem 5.82. Is every finite set compact? Justify your assertion.

The next theorem says that every nonempty compact set contains its greatest lower bound and its least upper bound. That is, every nonempty compact set attains a minimum and a maximum value.

Theorem 5.83. *If K is a nonempty compact subset of $\mathbb{R}$, then* $\sup(K), \inf(K) \in K$.

Definition 5.84. A set $A \subseteq \mathbb{R}$ is **disconnected** if there exists two disjoint open sets U_1 and U_2 such that $A \cap U_1$ and $A \cap U_2$ are nonempty but $A \subseteq U_1 \cup U_2$ (equivalently, $A = (A \cap U_1) \cup (A \cap U_2)$). If a set is not disconnected, then we say that it is **connected**.

In other words, a set is disconnected if it can be partitioned into two nonempty subsets such that each subset does not contain points of the other and does not contain any accumulation points of the other. Showing that a set is disconnected is generally easier than showing a set is connected. To prove that a set is disconnected, you simply need to exhibit two open sets with the necessary properties. However, to prove that a set is connected, you need to prove that no such pair of open sets exists.

Problem 5.85. Determine whether each of the sets in Problem 5.81 is is connected or disconnected. Briefly justify your assertions.

Theorem 5.86. *If $a \in \mathbb{R}$, then $\{a\}$ is connected.*

The proof of the next theorem is harder than you might expect. Consider a proof by contradiction and try to make use of the Completeness Axiom.

Theorem 5.87. *Every closed interval $[a, b]$ is connected.*

It turns out that every connected set in $\mathbb{R}$ is either a singleton or an interval. We have not officially proved this claim, but we do have the tools to do so. Feel free to try your hand at proving this fact.

> If you learn how to learn, it's the ultimate meta skill and I believe you can learn how to be healthy, you can learn how to be fit, you can learn how to be happy, you can learn how to have good relationships, you can learn how to be successful. These are all things that can be learned. So if you can learn that is a trump card, it's an ace, it's a joker, it's a wild card. You can trade it for any other skill.
>
> Naval Ravikant, entrepreneur & investor

> A mathematician, like a painter or a poet, is a maker of patterns. If his patterns are more permanent than theirs, it is because they are made with ideas.
>
> G.H. Hardy, mathematician

6 Three Famous Theorems

In the last few chapters, we have encountered all of the major proof techniques one needs in mathematics and enhanced our proof-writing skills. In this chapter, we put these techniques and skills to work to prove three famous theorems, as well as numerous intermediate results along the way. All of these theorems are ones you are likely familiar with from grade school, but perhaps these facts were never rigorously justified for you.

In the first section, we develop all of the concepts necessary to state and then prove the **Fundamental Theorem of Arithmetic** (Theorem 6.17), which you may not recognize by name. The Fundamental Theorem of Arithmetic states that every natural number greater than 1 is the product of a unique combination of prime numbers. To prove the Fundamental Theorem of Arithmetic, we will need to make use of the **Division Algorithm** (Theorem 6.7), which in turn utilizes the Well-Ordering Principle (Theorem 4.38). In the second section, we prove that $\sqrt{2}$ is irrational, which settles a claim made in Section 5.1. In the final section, we prove that there are infinitely many primes.

6.1 The Fundamental Theorem of Arithmetic

The goal of this section is to prove The Fundamental Theorem of Arithmetic. The Fundamental Theorem of Arithmetic (sometimes called the Unique Factorization Theorem) states that every natural number greater than 1 is either prime or is the product of prime numbers, where this product is unique up to the order of the factors. For example, the natural number 12 has prime factorization $2^2 \cdot 3$, where the order in which we write the prime factors (i.e., 2, 2, and 3) is irrelevant. That is, $2^2 \cdot 3$, $2 \cdot 3 \cdot 2$, and $3 \cdot 2^2$ are all the same prime factorization of 12. The requirement that the factors be prime is necessary since factorizations containing composite numbers may not be unique. For example, $12 = 2 \cdot 6$ and $12 = 3 \cdot 4$, but these factorizations into composite numbers are distinct. We have just thrown around a few fancy terms; we should make sure we understand their precise meaning.

Definition 6.1. Let $n \in \mathbb{Z}$.

(a) If $a \in \mathbb{Z}$ such that a divides n, then we say that a is a **factor** of n.

(b) If $n \in \mathbb{N}$ such that n has exactly two distinct positive factors (namely, 1 and n itself), then n is called **prime**.

(c) If $n > 1$ such that n is not prime, then n is called **composite**.

Problem 6.2. According to our definition, is 1 a prime number or composite number? Explain your answer. You may have heard prime numbers defined as something like, "a prime number is a natural number that is only divisible by 1 and itself." Does this definition agree with the one above?

The upshot is that according to our definition, 1 is neither prime nor composite. However, throughout history, this has not always been the case. There were times when and mathematicians for whom the number one was considered prime. Whether 1 is prime or not is a matter of definition, and hence a matter of choice. There are compelling reasons—that we will not elaborate on here—why 1 is intentionally excluded from being prime. However, if you would like to learn more, check out the excellent article "What is the Smallest Prime?" by Chris Caldwell and Yeng Xiong.

Problem 6.3. List the first 10 prime numbers.

Problem 6.4. Prove or provide a counterexample: For all $n \in \mathbb{N}$, if $4^n - 1$ is prime, then n is odd.

Problem 6.5. Prove or provide a counterexample: For all $n \in \mathbb{N}$, $n^2 - n + 11$ is prime.

The next result makes up half of the Fundamental Theorem of Arithmetic. We provide a substantial hint for its proof. Let S be the set of natural numbers for which the theorem fails. For sake of a contradiction, assume $S \neq \emptyset$. By the Well-Ordering Principle (Theorem 4.38), S contains a least element, say n. Then n cannot be prime since this would satisfy the theorem. So, it must be the case that n has a divisor other than 1 and itself. This implies that there exists natural numbers a and b greater than 1 such that $n = ab$. Since n was our smallest counterexample, what can you conclude about both a and b? Use this information to derive a counterexample for n.

Theorem 6.6. *If n is a natural number greater than 1, then n can be expressed as a product of primes. That is, we can write*

$$n = p_1 p_2 \cdots p_k,$$

where each of $p_1, p_2, \ldots, p_k$ is a prime number (not necessarily distinct).

Theorem 6.6 states that we can write every natural number greater than 1 as a product of primes, but it does not say that the primes and the number of times each prime appears are unique. To prove uniqueness, we will need **Euclid's Lemma** (Theorem 6.15). To prove Euclid's Lemma, we will utilize a special case of **Bézout's Lemma** (Theorem 6.13), the proof of which relies on the following result, known as the Division Algorithm. We include the proof of the Division Algorithm below, which makes use of the Well-Ordering Principle (Theorem 4.38).

Theorem 6.7 (Division Algorithm). *If $n, d \in \mathbb{Z}$ such that $d > 0$, then there exists unique $q, r \in \mathbb{Z}$ such that $n = dq + r$ with $0 \leq r < d$.*

Proof. Let $n, d \in \mathbb{Z}$ such that $d > 0$ such that $n > 0$. We have two tasks. First, we need to show that q and r exist, and then we need to show that both are unique.

If $d = 1$, it is clear that we can take $q = n$ and $r = 0$, so that $n = 1 \cdot n + 0 = dq + r$, as desired. Now, assume that $d > 1$ and define

$$S := \{n - dk \mid k \in \mathbb{Z} \text{ and } n - dk \geq 0\}.$$

If we can show that $S \neq \emptyset$, then we can apply the Well-Ordering Principle (Theorem 4.38) to conclude that S has a least element of S. This least element will be the remainder r we are looking for. There are two cases.

First, suppose $n \geq 0$. If we take $k = 0$, then we get $n - dk = n - d \cdot 0 = n \geq 0$, which shows that $n \in S$.

Now, suppose $n < 0$. In this case, we can take $k = n$, so that $n - dk = n - dn = n(1 - d)$. Since $n < 0$ and $d > 1$, $n(1 - d) > 0$. This shows that $n - dn \in S$.

We have shown that $S \neq \emptyset$, and so S contains a least element $r = n - dq$ for some $q \in \mathbb{Z}$. Then $n = dq + r$ with $r \geq 0$. For sake of a contradiction, assume $r \geq d$. This implies that there exists $r' \in \mathbb{Z}$ such that $r = d + r'$ and $0 \leq r' < r$. But then we see that

$$n = dq + r = dq + d + r' = d(q + 1) + r'.$$

This implies that $r' = n - d(q+1)$. Since $0 \leq r' < r$, we have produced an element of S that is smaller than r. This contradicts the fact that r is the least element of S, and so $r < d$.

It remains to show that q and r are unique. Suppose $q_1, q_2, r_1, r_2 \in \mathbb{Z}$ such that $n = dq_1 + r_1$ and $n = dq_2 + r_2$ and $0 \leq r_1, r_2 < d$. Without loss of generality, suppose $r_2 \geq r_1$, so that $0 \leq r_2 - r_1 < d$. Since $dq_1 + r_1 = dq_2 + r_2$, we see that

$r_2 - r_1 = d(q_1 - q_2)$. But then d divides $r_2 - r_1$. If $r_2 - r_1 > 0$, then by Theorem 2.56, it must be the case that $r_2 - r_1 \geq d$. However, we know $0 \leq r_2 - r_1 < d$, and so we must have $r_2 - r_1 = 0$. Therefore, $r_1 = r_2$, which in turn implies $q_1 = q_2$. We have shown that q and r are unique. □

In the Division Algorithm, we call n the **dividend**, d the **divisor**, q the **quotient**, and r the **remainder**. It is worth pointing out that the Division Algorithm holds more generally where the divisor d is not required to be positive. In this case, we must replace $0 \leq r < n$ with $0 \leq r < |n|$.

Contrary to its name, our statement of the Division Algorithm is not actually an algorithm, but this is the theorem's traditional name. However, there is an algorithm buried in this theorem. If n is nonnegative, repeatedly subtract d from n until we obtain an integer value that lies between 0 (inclusive) and d (exclusive). The resulting value is the remainder r while the number of times that d is subtracted is the quotient q. On the other hand, if n is negative, repeatedly add d to n until we obtain an integer value that lies between 0 (inclusive) and d (exclusive). Again, the resulting value is r. However, in this case, we take $-q$ to be the number of times that d is added, so that q (a negative value) is the quotient.

Problem 6.8. Suppose $n = 27$ and $d = 5$. Find the quotient and remainder that are guaranteed to exist by the Division Algorithm. That is, find the unique $q, r \in \mathbb{Z}$ such that $0 \leq r < n$ and $n = dq + r$.

It is a little trickier to determine q and r when n is negative.

Problem 6.9. Suppose $n = -26$ and $d = 3$. Find the quotient and remainder that are guaranteed to exist by the Division Algorithm. That is, find the unique $q, r \in \mathbb{Z}$ such that $0 \leq r < n$ and $n = dq + r$.

It is useful to have some additional terminology.

Definition 6.10. Let $m, n \in \mathbb{Z}$ such that at least one of m or n is nonzero. The **greatest common divisor** (gcd) of m and n, denoted $\gcd(m, n)$, is the largest positive integer that divides both m and n. If $\gcd(m, n) = 1$, we say that m and n are **relatively prime**.

Problem 6.11. Find $\gcd(54, 72)$.

Problem 6.12. Provide an example of two natural numbers that are relatively prime.

The next result is a special case of a theorem known as Bézout's Lemma (or Bézout's Identity). Ultimately, we will need this theorem to prove Euclid's

Lemma (Theorem 6.15), which we then use to prove uniqueness for the Fundamental Theorem of Arithmetic (Theorem 6.17). To prove our special case of Bézout's Lemma, consider the set $S := \{ps + at > 0 \mid s, t \in \mathbb{Z}\}$. First, observe that $p \in S$ (choose $s = 1$ and $t = 0$). It follows that S is nonempty. By the Well-Ordering Principle (Theorem 4.38), S contains a least element, say d. This implies that there exists $s_1, t_1 \in \mathbb{Z}$ such that $d = ps_1 + at_1$. Our goal is to show that $d = 1$. Now, choose $m \in S$. Then there exists $s_2, t_2 \in \mathbb{Z}$ such that $m = ps_2 + at_2$. By the definition of d, we know $d \leq m$. By the Division Algorithm, there exists unique $q, r \in \mathbb{N} \cup \{0\}$ such that $m = qd + r$ with $0 \leq r < d$. Now, solve for r and then replace m and d with $ps_1 + at_1$ and $ps_2 + at_2$, respectively. You should end up with an expression for r involving p, a, s_1, s_2, t_1, and t_2. Next, rearrange this expression to obtain r as a linear combination of p and a (i.e., a sum of a multiple of p and a multiple of a). What does the minimality of d imply about r? You should be able to conclude that m is a multiple of d. That is, every element of S is a multiple of d. However, recall that $p \in S$, p is prime, and p and a are relatively prime. What can you conclude about d?

Theorem 6.13 (Special Case of Bézout's Lemma). *If $p, a \in \mathbb{Z}$ such that p is prime and p and a are relatively prime, then there exists $s, t \in \mathbb{Z}$ such that $ps + at = 1$.*

Problem 6.14. Consider the natural numbers 2 and 7, which happen to be relatively prime. Find integers s and t guaranteed to exist according to Theorem 6.13. That is, find $s, t \in \mathbb{Z}$ such that $2s + 7t = 1$.

The following theorem is known as Euclid's Lemma. Note that if p divides a, the conclusion is certainly true. So, assume otherwise. That is, assume that p does not divide a, so that p and a are relatively prime. Apply Theorem 6.13 to p and a and then multiply the resulting equation by b. Try to conclude that p divides b.

Theorem 6.15 (Euclid's Lemma). *Assume that p is prime. If p divides ab, where $a, b \in \mathbb{N}$, then either p divides a or p divides b.*

In Euclid's Lemma, it is crucial that p is prime as illustrated by the next problem.

Problem 6.16. Provide an example of integers a, b, d such that d divides ab yet d does not divide a and d does not divide b.

Alright, we are finally ready to tackle the proof of the Fundamental Theorem of Arithmetic. Let n be a natural number greater than 1. By Theorem 6.6, we know that n can be expressed as a product of primes. All that remains is to prove that this product is unique (up to the order in which they appear). For sake of a contradiction, suppose $p_1 p_2 \cdots p_k$ and $q_1 q_2 \cdots q_l$ are both prime factorizations of n. Your goal is to prove that $k = l$ and that each p_i is equal to some q_j. Make repeated use of Euclid's Lemma.

Theorem 6.17 (Fundamental Theorem of Arithmetic). *Every natural number greater than 1 can be expressed uniquely (up to the order in which they appear) as the product of one or more primes.*

The Fundamental Theorem of Arithmetic is one of the many reasons why 1 is not considered a prime number. If 1 were prime, prime factorizations would not be unique.

> Any creative endeavor is built on the ash heap of failure.
>
> Michael Starbird, mathematician

6.2 The Irrationality of $\sqrt{2}$

In this section we will prove one of the oldest and most important theorems in mathematics: $\sqrt{2}$ is irrational (see Theorem 6.19). First, we need to know what this means.

Definition 6.18. Let $r \in \mathbb{R}$.

(a) We say that r is **rational** if $r = \dfrac{m}{n}$, where $m, n \in \mathbb{Z}$ and $n \neq 0$.

(b) In contrast, we say that r is **irrational** if it is not rational.

The Pythagoreans were an ancient secret society that followed their spiritual leader: Pythagoras of Samos (c. 570–495 BCE). The Pythagoreans believed that the way to spiritual fulfillment and to an understanding of the universe was through the study of mathematics. They believed that all of mathematics, music, and astronomy could be described via whole numbers and their ratios. In modern mathematical terms they believed that all numbers are rational. Attributed to Pythagoras is the saying, "Beatitude is the knowledge of the perfection of the numbers of the soul." And their motto was "All is number."

Thus they were stunned when one of their own—Hippasus of Metapontum (c. 5th century BCE)—discovered that the side and the diagonal of a square are incommensurable. That is, the ratio of the length of the diagonal to the length of the side is irrational. Indeed, if the side of the square has length a, then the diagonal will have length $a\sqrt{2}$; the ratio is $\sqrt{2}$ (see Figure 6.1).

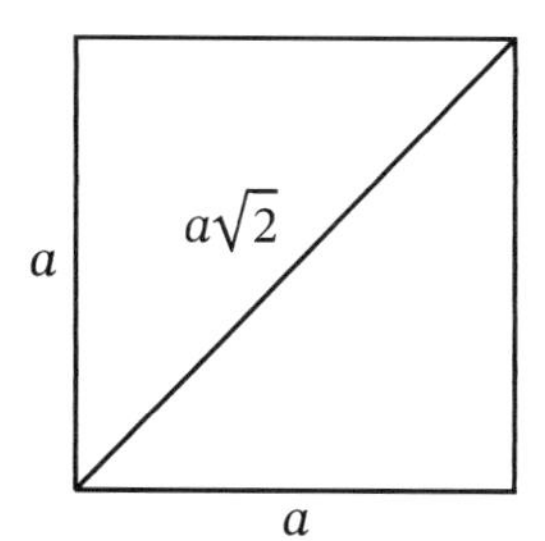

Figure 6.1. The side and diagonal of a square are incommensurable.

In Section 5.1, we took for granted that $\sqrt{2}$ was irrational. We now prove this fact. Consider using a proof by contradiction. Suppose that there exist $m, n \in \mathbb{Z}$ such that $n \neq 0$ and $\sqrt{2} = \frac{m}{n}$. Are there an odd or even number of factors of 2 on each side of this equation? Does your conclusion violate the Fundamental Theorem of Arithmetic (Theorem 6.17)?

Theorem 6.19. *The real number $\sqrt{2}$ is irrational.*

As one might expect, the Pythagoreans were unhappy with this discovery. Legend says that Hippasus was expelled from the Pythagoreans and was perhaps drowned at sea. Ironically, this result, which angered the Pythagoreans so much, is probably their greatest contribution to mathematics: the discovery of irrational numbers.

See if you can generalize the technique in the proof of Theorem 6.19 to prove the next two theorems.

Theorem 6.20. *Let p be a prime number. Then $\sqrt{p}$ is irrational.*

Theorem 6.21. *Let p and q be distinct primes. Then $\sqrt{pq}$ is irrational.*

Problem 6.22. State a generalization of Theorem 6.21 and briefly describe how its proof would go. Be as general as possible.

It is important to point out that not every positive irrational number is equal to the square root of some natural number. For example, π is irrational, but is not equal to the square root of a natural number.

> Getting better is not pretty. To get good we have to be down to struggle, seek out challenges, make some mistakes, to train ugly.
>
> Trevor Ragan, `thelearnerlab.com`

6.3 The Infinitude of Primes

The highlight of this section is Theorem 6.25, which states that there are infinitely many primes. The first known proof of this theorem is in Euclid's *Elements* (c. 300 BCE). Euclid stated it as follows:

> **Proposition IX.20.** *Prime numbers are more than any assigned multitude of prime numbers.*

There are a few interesting observations to make about Euclid's proposition and his proof. First, notice that the statement of the theorem does not contain the word "infinity." The Greek's were skittish about the idea of infinity. Thus, he proved that there were more primes than any given finite number. Today we would say that there are infinitely many. In fact, Euclid proved that there are more than *three* primes and concluded that there were more than any finite number. While such a proof is not considered valid in the modern era, we can forgive Euclid for this less-than-rigorous proof; in fact, it is easy to turn his proof into the general one that you will give below. Lastly, Euclid's proof was geometric. He was viewing his numbers as line segments with integral length. The modern concept of number was not developed yet.

Prior to tackling a proof of Theorem 6.25, we need to prove a couple of preliminary results. The proof of the first result is provided for you.

Theorem 6.23. *The only natural number that divides* 1 *is* 1.

Proof. Let m be a natural number that divides 1. We know that $m \geq 1$ because 1 is the smallest positive integer. Since m divides 1, there exists $k \in \mathbb{N}$ such that $1 = mk$. Since $k \geq 1$, we see that $mk \geq m$. But $1 = mk$, and so $1 \geq m$. Thus, we have $1 \leq m \leq 1$, which implies that $m = 1$, as desired. □

For the next theorem, try utilizing a proof by contradiction together with Theorem 6.23.

Theorem 6.24. *Let p be a prime number and let $n \in \mathbb{Z}$. If p divides n, then p does not divide $n + 1$.*

We are now ready to prove the following important theorem. Use a proof by contradiction. In particular, assume that there are finitely many primes, say $p_1, p_2, \ldots, p_k$. Consider the product of all of them and then add 1.

Theorem 6.25. *There are infinitely many prime numbers.*

We conclude this chapter with a fun problem involving prime numbers. This problem comes from David Richeson (Dickinson College).

Problem 6.26. Start with the first n prime numbers, $p_1, \ldots, p_n$. Divide them into two sets. Let a be the product of the primes in one set and let b be the product of the primes in the other set. Assume the product is 1 if the set is empty. For example, if $n = 5$, we could have $\{2, 7\}$ and $\{3, 5, 11\}$, and so $a = 14$ and $b = 165$. In general, what can we conclude about $a + b$ and $a - b$? Form a conjecture and then prove it.

It does not matter how slowly you go as long as you do not stop.

Confucius, philosopher

The impediment to action advances action.
What stands in the way becomes the way.

Marcus Aurelius, Roman emperor

7 Relations and Partitions

While there is no agreed upon universal definition of mathematics, one could argue that mathematics focuses on the study of patterns and relationships. Certain types of relationships occur over and over in mathematics. One way of formalizing the abstract nature and structure of these relationships is with the notion of relations. In Chapter 8, we will see that a function is a special type of relation.

7.1 Relations

Recall from Section 3.5 that the Cartesian product of two sets A and B, written $A \times B$, is the set of all ordered pairs (a, b), where $a \in A$ and $b \in B$. That is, $A \times B = \{(a, b) \mid a \in A, b \in B\}$.

Definition 7.1. Let A and B be sets. A **relation** R **from** A **to** B is a subset of $A \times B$. If R is a relation from A to B and $(a, b) \in R$, then we say that a **is related to** b and we may write $\boxed{aRb}$ in place of $(a, b) \in R$. If R is a relation from A to the same set A, then we say that R is a **relation on** A.

Example 7.2. The set $\mathbb{N} \times \mathbb{R}$ from Problem 3.55 is an example of a relation on $\mathbb{R}$ since $\mathbb{N} \times \mathbb{R}$ is a subset of $\mathbb{R} \times \mathbb{R}$.

It is important to notice that the order in which we write things for relations matters. In particular, if R is a relation from A to B and aRb, then it may or may not be the case that bRa.

Example 7.3. If $A = \{a, b, c, d, e\}$ and $B = \{1, 2, 3, 4\}$, then the set of ordered pairs

$$R = \{(a, 1), (a, 2), (a, 4), (c, 2), (d, 2), (e, 2), (e, 4)\}$$

is an example of a relation from A to B. In this case, we could write $(c, 2) \in R$ or $cR2$. We could also say that a is related to 1, 2, and 4.

Example 7.4. As in the previous example, let $A = \{a, b, c, d, e\}$. One possible relation on A is given by

$$R = \{(a, a), (a, b), (a, c), (b, b), (b, a), (b, c), (c, d), (c, e), (d, d), (d, a), (d, c), (e, a)\}.$$

Example 7.5. Consider the set of accounts A on the social media platform Twitter. On Twitter, each account has a set of accounts that they follow. We can model this situation mathematically using a relation on A. Define T on A via xTy if x follows y on Twitter. As a set

$$T = \{(x, y) \in A \times A \mid x \text{ follows } y \text{ on Twitter}\}.$$

Example 7.6. You are already familiar with many relations. For example, $=$, $\leq$, and $<$ are each examples of relations on the real numbers. We could say that $(3, \pi)$ is in the relation $\leq$ and the relation $<$ since $3 \leq \pi$ and $3 < \pi$. However, $(3, \pi)$ is not in the relation $=$ since $3 \neq \pi$. Also, notice that order matters for the relations $\leq$ and $<$ yet does not for $=$. For example, $(-\sqrt{2}, 4)$ is in the relation $\leq$ while $(4, -\sqrt{2})$ is not.

Example 7.7. Define the relation S from $\{-1, 1\}$ to $\mathbb{Z}$ via $1Sx$ if x is even and $-1Sx$ if x is odd. That is, 1 is related to all even integers and -1 is related to all odd integers.

Example 7.8. Let A be any set. Since $\emptyset \subseteq A \times A$, the empty set forms a relation on A. This relation is called the **empty relation** on A.

Relations can be represented using digraphs. A **digraph** (short for **directed graph**) is a discrete graph that consists of a set of vertices connected by edges, where the edges have a direction associated with them. If R is a relation from A to B, then the elements of A and B are the vertices of the digraph and there is a directed edge from $a \in A$ to $b \in B$ if (a, b) is in the relation R (i.e., aRb). We can visually represent digraphs by using dots to represent the vertices and arrows to represent directed edges. We will not make a distinction between a digraph and its visual representation. Utilizing a digraph to represent a relation may be impractical if there is a large number of vertices or directed edges.

Example 7.9. Consider the relation given in Example 7.3. The corresponding digraph is depicted in Figure 7.1. Notice that we have placed the vertices corresponding to elements of A on the left and the elements of B on the right. This is standard practice, but what really matters is the edge connections not how the vertices are placed on the page.

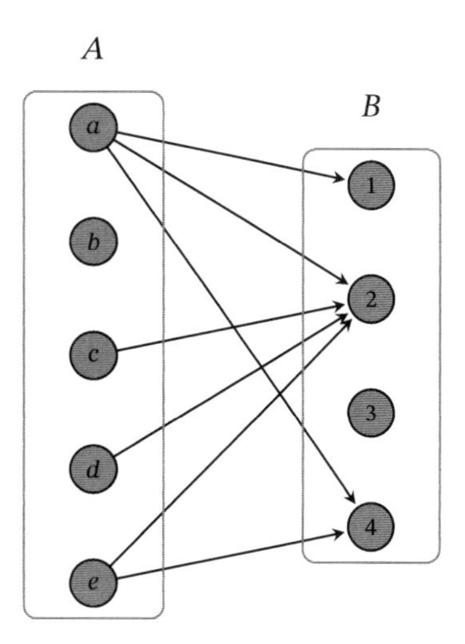

Figure 7.1. Digraph for a relation from $A = \{a, b, c, d, e\}$ to $B = \{1, 2, 3, 4\}$.

Problem 7.10. Let $A = \{1, 2, 3, 4, 5, 6\}$ and $B = \{1, 2, 3, 4\}$ and define D from A to B via $(a, b) \in D$ if $a - b$ is divisible by 2. List the ordered pairs in D and draw the corresponding digraph.

If R is a relation on A (i.e., a relation from A to A), then we can simplify the structure of the digraph by only utilizing one copy of A for the vertices. In this case, we may have directed edges that point from a vertex to itself. When drawing digraphs for a relation on a set, we will default to this simplified digraph (like the one depicted in Figure 7.2(b)).

Example 7.11. Figure 7.2(a) represents the relation of Example 7.4 as a digraph from A to A while the digraph in Figure 7.2(b) provides a streamlined representation of the same relation that uses the elements in A only once instead of twice.

Problem 7.12. Let $A = \{1, 2, 3, 4, 5, 6\}$ and define $\mid$ on A via $x|y$ if x divides y. List the ordered pairs in $\mid$ and draw the corresponding digraph.

Problem 7.13. Let $A = \{a, b, c, d\}$ and define R on A via

$$R = \{(a, a), (a, b), (a, c), (b, b), (b, a), (b, c), (c, c), (c, a), (c, b), (d, d)\}.$$

(a) Draw the digraph for R.

(b) Draw the digraph for the empty relation on A.

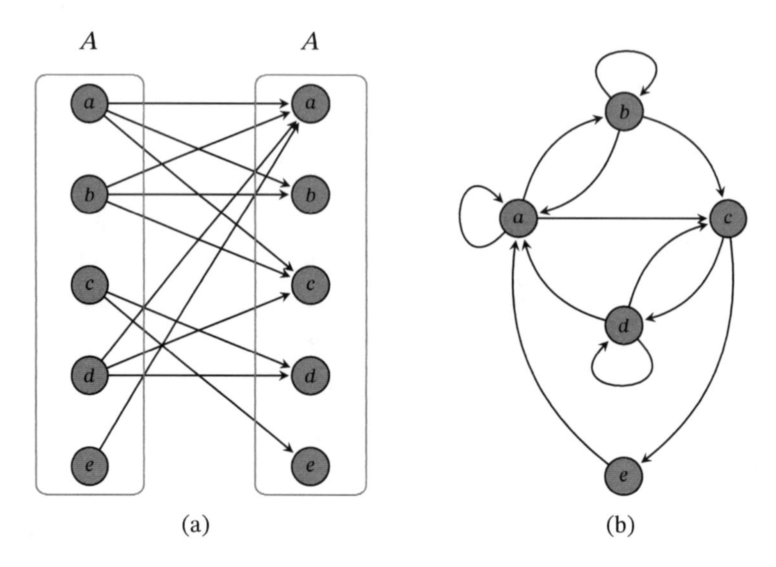

Figure 7.2. Two variations of digraphs for a relation on $A = \{a, b, c, d, e\}$.

We can also visually represent a relation by plotting the points in the relation. In particular, if R is a relation from A to B and aRb, we can plot all points (a, b) that satisfy aRb in two dimensions, where we interpret the set A to be the horizontal axis and B to be the vertical axis. We will refer to this visual representation of a relation as the **graph** of the relation.

Example 7.14. When we write $x^2 + y^2 = 1$, we are implicitly defining a relation. In particular, the relation is the set of ordered pairs (x, y) satisfying $x^2 + y^2 = 1$, namely $\{(x, y) \in \mathbb{R}^2 \mid x^2 + y^2 = 1\}$. The graph of this relation in $\mathbb{R}^2$ is the unit circle centered at the origin in the plane as shown in Figure 7.3.

Problem 7.15. For each of the following, draw a portion of the graph that represents the relation as a subset of $\mathbb{R}^2$.

(a) $\{(x, y) \in \mathbb{R}^2 \mid y = x^2\}$

(b) $\{(x, y) \in \mathbb{Z}^2 \mid y = x^2\}$

(c) $\{(x, y) \in \mathbb{R}^2 \mid y^2 = x\}$

(d) $\{(x, y) \in \mathbb{N} \times \mathbb{R} \mid y^2 = x\}$

Problem 7.16. Draw a portion of the graph that represents the relation $\leq$ on $\mathbb{R}$.

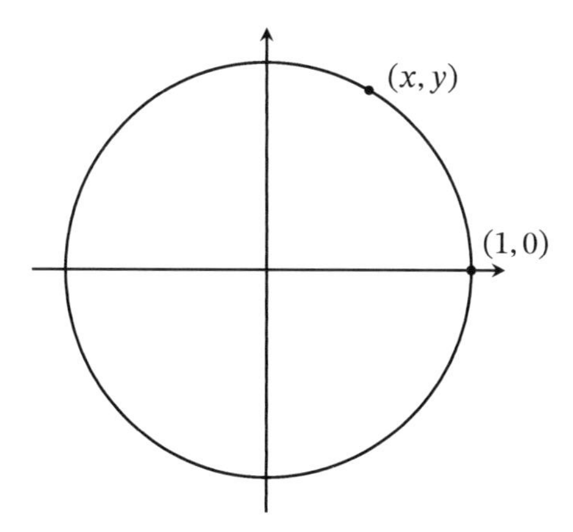

Figure 7.3. Graph of the relation determined by $x^2 + y^2 = 1$.

For a relation on a set, it is natural to consider the collection of elements that a given element is related to. For example, a user's "Following List" on Twitter is the set of accounts on Twitter that the user is following.

Definition 7.17. Let R be a relation on a set A. For each $a \in A$, we define the **set of relatives of** a **with respect to** R via

$$\boxed{\text{rel}(a, R) := \{b \in A \mid aRb\}}.$$

We also define the **collection of the sets of relatives with respect to** R by

$$\boxed{\text{Rel}(R) := \{\text{rel}(a) \mid a \in A\}}.$$

If R is clear from the context, we will usually write $\boxed{\text{rel}(a)}$ in place of $\text{rel}(a, R)$. In terms of digraphs, $\text{rel}(a)$ is the collection of vertices that have a directed edge pointing towards them from the vertex labeled by a. In graph theory, this collection of vertices is called the **out neighborhood** of a and each such vertex is called an **out neighbor**. Notice that $\text{Rel}(R)$ is a set of sets. In particular, an element in $\text{Rel}(R)$ is a subset of A—equivalently, an element of $\mathcal{P}(A)$.

Example 7.18. Consider the relation given in Example 7.4. By inspecting the ordered pairs in R or by looking at the digraph in Figure 7.2(b), we see that $\text{rel}(a) = \{a, b, c\}$, $\text{rel}(b) = \{a, b, c\}$, $\text{rel}(c) = \{d, e\}$, $\text{rel}(d) = \{a, c, d\}$, $\text{rel}(e) = \{a\}$, so that $\text{Rel}(R) = \{\{a, b, c\}, \{d, e\}, \{a, c, d\}, \{a\}\}$.

Problem 7.19. Consider the relation given in Problem 7.13(a). Find $\text{Rel}(R)$ by determining $\text{rel}(x)$ for each $x \in A$.

Problem 7.20. Describe the collection of the sets of relatives with respect to the empty relation from Problem 7.13(b).

Problem 7.21. Let P denote the set of all people with accounts on Facebook and define the relation F on P via xFy if x is friends with y. Describe rel(Maria), where Maria is the name of a specific Facebook user. What is $\text{Rel}(F)$?

Problem 7.22. Define the relation $\equiv_5$ on $\mathbb{Z}$ via $a \equiv_5 b$ if $a - b$ is divisible by 5. Find $\text{rel}(1)$, $\text{rel}(2)$, and $\text{rel}(6)$. How many distinct sets are in $\text{Rel}(\equiv_5)$? List the distinct sets in $\text{Rel}(\equiv_5)$.

Problem 7.23. Consider the relation $\leq$ on $\mathbb{R}$. If $x \in \mathbb{R}$, what is $\text{rel}(x)$?

Problem 7.24. Suppose R is a relation on $A = \{1, 2, 3, 4, 5\}$ such that $\text{rel}(1) = \{1, 3, 4\}$, $\text{rel}(2) = \{4\}$, $\text{rel}(3) = \{3, 4, 5\}$, $\text{rel}(4) = \{1, 2\}$, and $\text{rel}(5) = \emptyset$. List the ordered pairs in R and draw the corresponding digraph.

We will now examine three important properties that a relation on a set may or may not possess.

Definition 7.25. Let R be a relation on a set A.

(a) The relation R is **reflexive** if for all $a \in A$, aRa.

(b) The relation R is **symmetric** if for all $a, b \in A$, if aRb, then bRa.

(c) The relation R is **transitive** if for all $a, b, c \in A$, if aRb and bRc, then aRc.

Example 7.26. Here are a few examples that illustrate the concepts in the previous definition.

(a) The relation $=$ on $\mathbb{R}$ is reflexive, symmetric, and transitive.

(b) The relation $\leq$ is reflexive and transitive on $\mathbb{R}$, but not symmetric. However, notice that $<$ is transitive on $\mathbb{R}$, but neither symmetric nor reflexive.

(c) If S is a set, then $\subseteq$ on $\mathcal{P}(S)$ is reflexive and transitive, but not symmetric.

Problem 7.27. Determine whether the relations given in each of the following is reflexive, symmetric, or transitive.

(a) Example 7.4

(b) Problem 7.13

Problem 7.28. Suppose R is a relation on a set A.

(a) Explain what it means for R to *not* be reflexive.

(b) Explain what it means for R to *not* be symmetric.

(c) Explain what it means for R to *not* be transitive.

Problem 7.29. Let $A = \{a, b, c, d, e\}$.

(a) Define a relation R on A that is reflexive but not symmetric nor transitive.

(b) Define a relation S on A that is symmetric but not reflexive nor transitive.

(c) Define a relation T on A that is transitive but not reflexive nor symmetric.

Problem 7.30. Given a relation R on a finite set A, describe what each of reflexive, symmetric, and transitive look like in terms of a digraph. That is, draw pictures that represent each of reflexive, symmetric, and transitive. One thing to keep in mind is that the elements used in the definitions of symmetric and transitive do not have to be distinct. So, you might need to consider multiple cases.

Below, we provide skeleton proofs for proving that a relation is reflexive, symmetric, or transitive. Notice that the skeleton proof for proving that a relation is reflexive is a special case of Skeleton Proof 2.81. Similarly, the skeleton proofs involving symmetric and transitive are both special cases of Skeleton Proof 2.82. It is important to point out that every relation on the empty set is vacuously reflexive, symmetric, and transitive. In the skeleton proofs below, we are implicitly assuming that the set in question is nonempty. In some circumstances, it may be necessary to mention the possibility of the empty set.

Skeleton Proof 7.31 (Proof that a relation is reflexive)**.** Here is the general structure for proving that a relation is reflexive.

Proof. Assume R is a relation on A defined by [or satisfying]... [*Use the given definition (or describe the given property) of* R]. Let $a \in A$.

... [*Use the definition (or property) of* R *to verify that* aRa] ...

Therefore, the relation R is reflexive on A. □

Skeleton Proof 7.32 (Proof that a relation is symmetric). Here is the general structure for proving that a relation is symmetric.

> *Proof.* Assume R is a relation on A defined by (or satisfying)... [*Use the given definition (or describe the given property) of R*]. Let $a, b \in A$ and suppose aRb.
>
> ... [*Use assumption that aRb with definition (or property) of R to verify that bRa*] ...
>
> Therefore, the relation R is symmetric on A. □

Skeleton Proof 7.33 (Proof that a relation is transitive). Here is the general structure for proving that a relation is transitive.

> *Proof.* Assume R is a relation on A defined by (or satisfying)... [*Use the given definition (or describe the given property) of R*]. Let $a, b, c \in A$ and suppose aRb and bRc.
>
> ... [*Use assumption that aRb and bRc with definition (or property) of R to verify that aRc*] ...
>
> Therefore, the relation R is transitive on A. □

Problem 7.34. Determine whether each of the following relations is reflexive, symmetric, or transitive. In each case, you should either provide a specific counterexample or a proof.

(a) Consider the relation T described in Example 7.5.

(b) Consider the relation F described in Problem 7.21.

(c) Consider the relation $\equiv_5$ described in Problem 7.22.

(d) Let P be the set of all people and define H via xHy if x and y have the same height.

(e) Let P be the set of all people and define T via xTy if x is taller than y.

(f) Consider the relation "divides" on $\mathbb{N}$.

(g) Let L be the set of lines and define $||$ via $l_1 || l_2$ if l_1 is parallel to l_2.

(h) Let $C[0, 1]$ be the set of continuous functions on $[0, 1]$. Define $f \sim g$ if

$$\int_0^1 |f(x)|\, dx = \int_0^1 |g(x)|\, dx.$$

(i) Define R on $\mathbb{N}$ via nRm if $n + m$ is even.

(j) Define D on $\mathbb{R}$ via $(x, y) \in D$ if $x = 2y$.

(k) Define F on $\mathbb{Z} \times (\mathbb{Z} \setminus \{0\})$ via $(a, b)F(c, d)$ if $ad = bc$. Do you recognize this relation? Think about fractions.

(l) Define $\sim$ on $\mathbb{R}^2$ via $(x_1, y_1) \sim (x_2, y_2)$ if $x_1^2 + y_1^2 = x_2^2 + y_2^2$.

(m) Define S on $\mathbb{R}$ via xSy if $\lfloor x \rfloor = \lfloor y \rfloor$, where $\lfloor x \rfloor$ is the greatest integer less than or equal to x (e.g., $\lfloor \pi \rfloor = 3$, $\lfloor -1.5 \rfloor = -2$, and $\lfloor 4 \rfloor = 4$).

(n) Define C on $\mathbb{R}$ via xCy if $|x - y| < 1$.

Most of what we believe, we believe because it was told to us by someone we trusted. What I would like to suggest, however, is that if we rely too much on that kind of education, we could find in the end that we have never really learned anything.

Paul Wallace, physicist & theologian

7.2 Equivalence Relations

As we have seen in the previous section, the notions of reflexive, symmetric, and transitive are independent of each other. That is, a relation may have some combination of these properties, possibly none of them and possibly all of them. However, we have a special name for when a relation satisfies all three properties.

Definition 7.35. Let $\sim$ be a relation on a set A. Then $\sim$ is called an **equivalence relation** on A if $\sim$ is reflexive, symmetric, and transitive.

The symbol "$\sim$" is usually pronounced as "twiddle" or "tilde" and the phrase "$a \sim b$" could be read as "a is related to b" or "a twiddles b".

Problem 7.36. Let $A = \{1, 2, 3, 4, 5, 6\}$ and define

$$R = \{(1,1), (1,6), (2,2), (2,3), (2,4), (3,3), (3,2), (3,4), (4,4), (4,2), (4,3), (5,5), (6,6), (6,1)\}.$$

Using R, complete each of the following.

(a) Draw the digraph for R.

(b) Determine whether R is an equivalence relation on A.

(c) Find $\mathrm{Rel}(R)$ by determining $\mathrm{rel}(x)$ for each $x \in A$.

Problem 7.37. Let $A = \{a, b, c, d, e\}$.

(a) Make up an equivalence relation $\sim$ on A by drawing a digraph such that a is not related to b and c is not related to b.

(b) Using your digraph, find $\mathrm{Rel}(\sim)$ by determining $\mathrm{rel}(x)$ for each $x \in A$.

Problem 7.38. Given a finite set A and an equivalence relation $\sim$ on A, describe what the corresponding digraph would have to look like.

Problem 7.39. Determine which relations given in Problem 7.34 are equivalence relations.

Problem 7.40. Let $\mathcal{T}$ be the set of all triangles and define $\sim$ on $\mathcal{T}$ via $T_1 \sim T_2$ if T_1 is similar to T_2. Determine whether $\sim$ is an equivalence relation on $\mathcal{T}$.

Problem 7.41. If possible, construct an equivalence relation on the empty set. If this is not possible, explain why.

Theorem 7.42. *Suppose $\sim$ is an equivalence relation on a set A and let $a, b \in A$. Then* $\mathrm{rel}(a) = \mathrm{rel}(b)$ *if and only if $a \sim b$.*

Theorem 7.43. *Suppose $\sim$ is an equivalence relation on a set A. Then*

(a) $\bigcup_{a \in A} \mathrm{rel}(a) = A$, *and*

(b) *For all $a, b \in A$, either* $\mathrm{rel}(a) = \mathrm{rel}(b)$ *or* $\mathrm{rel}(a) \cap \mathrm{rel}(b) = \emptyset$.

In light of Theorem 7.43, we have the following definition.

Definition 7.44. If $\sim$ is an equivalence relation on a set A, then for each $a \in A$, we refer to $\mathrm{rel}(a)$ as the **equivalence class** of a.

When $\sim$ is an equivalence relation on a set A, it is common to write each equivalence class $\mathrm{rel}(a)$ as $[a]$ (or sometimes $\overline{a}$). The element a inside the square brackets is called the **representative of the equivalence class** $[a]$. Theorem 7.42 implies that an equivalence class can be represented by any element of the equivalence class. For example, in Problem 7.36, we have $[1] = [6]$ since 1 and 6 are in the same equivalence class. The collection of equivalence classes $\mathrm{Rel}(\sim)$ is often denoted by $A/{\sim}$, which is read as "A modulo $\sim$" or "A mod $\sim$". The collection $A/{\sim}$ is sometimes referred to as the **quotient of** A **by** $\sim$.

Example 7.45. Let P denote the residents of a particular town and define $\sim$ on P via $a \sim b$ if a and b have the same last name. It is easily seen that this relation is reflexive, symmetric, and transitive, and hence $\sim$ is an equivalence relation on P. The equivalence classes correspond to collections of individuals with the same last name. For example, Maria Garcia, Anthony Garcia, and Ariana Garcia all belong to the same equivalence class. Any Garcia can be used as a representative for the corresponding equivalence class, so we can denote it as [Maria Garcia], for example. The collection $P/\sim$ consists of the various sets of people with the same last name. In particular, [Maria Garcia] $\in P/\sim$.

Example 7.46. The five distinct sets of relatives that you identified in Problem 7.22 are the equivalence classes for $\equiv_5$ on $\mathbb{Z}$. These equivalence classes are often called the **congruence classes modulo 5**.

The upshot of Theorem 7.43 is that given an equivalence relation, every element lives in exactly one equivalence class. In the next section, we will see that we can run this in reverse. That is, if we separate out the elements of a set so that every element is an element of exactly one subset, then this determines an equivalence relation.

Problem 7.47. If $\sim$ is an equivalence relation on a finite set A, describe $A/\sim$ in terms of the digraph corresponding to $\sim$.

Problem 7.48. For each of the equivalence relations you identified in Problem 7.39, succinctly describe the corresponding equivalence classes.

Problem 7.49. Suppose R and S are both equivalence relations on a set A. Is $R \cap S$ an equivalence relation on A? If so, prove it. Otherwise, provide a counterexample.

Problem 7.50. Suppose R and S are both equivalence relations on a set A. Is $R \cup S$ an equivalence relation on A? If so, prove it. Otherwise, provide a counterexample.

> Mathematics has beauty and romance. It's not a boring place to be, the mathematical world. It's an extraordinary place; it's worth spending time there.
>
> Marcus du Sautoy, mathematician

7.3 Partitions

Theorems 7.42 and 7.43 imply that if $\sim$ is an equivalence relation on a set A, then $\sim$ breaks A up into pairwise disjoint "chunks", where each chunk is some $[a]$ for $a \in A$. As you have probably already noticed, equivalence relations are intimately related to the following concept.

Definition 7.51. A collection Ω of subsets of a set A is said to be a **partition** of A if the elements of Ω satisfy:

(a) Each $X \in \Omega$ is nonempty,

(b) For all $X, Y \in \Omega$, $X \cap Y = \emptyset$ when $X \neq Y$, and

(c) $\bigcup_{X \in \Omega} X = A$.

That is, the elements of Ω are pairwise disjoint nonempty sets and their union is all of A. Each $X \in \Omega$ is called a **block** of the partition.

Example 7.52. Consider the equivalence relation $\sim$ on the set P described in Example 7.45. Recall that the equivalence classes correspond to collections of individuals with the same last name. Since each equivalence class is nonempty and each resident of the town belongs to exactly one equivalence class, the collection of equivalence classes forms a partition of P. That is, $P/\sim$ is a partition of P, where the blocks of the partition correspond to sets of residents with the same last name.

Example 7.53. Each of the following is an example of a partition of the set given in parentheses.

(a) Democrat, Republican, Independent, Green Party, Libertarian, etc. (set of registered voters)

(b) Freshman, sophomore, junior, senior (set of high school students)

(c) Evens, odds (set of integers)

(d) Rationals, irrationals (set of real numbers)

Example 7.54. Let $A = \{a, b, c, d, e, f\}$ and $\Omega = \{\{a\}, \{b, c, d\}, \{e, f\}\}$. Since the elements of Ω are pairwise disjoint nonempty subsets of A such that their union is all of A, Ω is a partition of A consisting of three blocks.

Problem 7.55. Consider the set A from Example 7.54.

(a) Find a partition of A consisting of four blocks.

(b) Find a collection of subsets of A that does *not* form a partition. See how many ways you can prevent your collection from being a partition.

Problem 7.56. For each of the following, find a partition of $\mathbb{Z}$ with the given properties.

(a) A partition of $\mathbb{Z}$ that consists of finitely many blocks, where each of the blocks is infinite.

(b) A partition of $\mathbb{Z}$ that consists of infinitely many blocks, where each of the blocks is finite.

(c) A partition of $\mathbb{Z}$ that consists of infinitely many blocks, where each of the blocks is infinite.

Problem 7.57. For each relation in Problem 7.34, determine whether the corresponding collection of the sets of relatives forms a partition of the given set.

Problem 7.58. Can we partition the empty set? If so, describe a partition. If not, explain why.

The next theorem spells out half of the close connection between partitions and equivalence relations. Theorem 7.73 yields the other half.

Theorem 7.59. *If $\sim$ is an equivalence relation on a nonempty set A, then $A/\sim$ forms a partition of A.*

Problem 7.60. In the previous theorem, why did we require A to be nonempty?

Problem 7.61. Consider the equivalence relation

$$\begin{aligned}\sim = \{&(1,1),(1,2),(2,1),(2,2),(3,3),\\ &(4,4),(4,5),(5,4),(5,5),(6,6),(5,6),(6,5),(4,6),(6,4)\}\end{aligned}$$

on the set $A = \{1,2,3,4,5,6\}$. Find the partition determined by $\text{Rel}(\sim)$.

It turns out that we can reverse the situation, as well. That is, given a partition, we can form an equivalence relation such that the equivalence classes correspond to the blocks of the partition. Before proving this, we need a definition.

Definition 7.62. Let A be a set and Ω any collection of subsets of A (not necessarily a partition). Define the relation $\boxed{R_\Omega}$ on A via $aR_\Omega b$ if there exists $X \in \Omega$ that contains both a and b. This relation is called the **relation on A associated to Ω**.

In other words, two elements are related exactly when they are in the same subset.

Problem 7.63. Let $A = \{a,b,c,d,e,f\}$ and let $\Omega = \{\{a,c\},\{b,c\},\{d,f\}\}$. List the ordered pairs in R_Ω and draw the corresponding digraph.

Problem 7.64. Let A and Ω be as in Example 7.54. List the ordered pairs in R_Ω and draw the corresponding digraph.

Problem 7.65. Consider Problem 7.24. Find the relation on A associated to $\text{Rel}(\sim)$ and compare with what you obtained for R in Problem 7.24.

Problem 7.66. Give an example of a set A and a collection Ω from $\mathcal{P}(A)$ such that the relation R_Ω is not reflexive.

Problem 7.67. Let $A = \{1, 2, 3, 4, 5, 6\}$ and $\Omega = \{\{1, 3, 4\}, \{2, 4\}, \{3, 4\}, \{6\}\}$.

(a) Is Ω a partition of A?

(b) Find R_Ω by listing ordered pairs or drawing a digraph.

(c) Is R_Ω an equivalence relation?

(d) Find $\text{Rel}(R_\Omega)$ (i.e., the collection of subsets of A determined by R_Ω). How are Ω and $\text{Rel}(R_\Omega)$ related?

Theorem 7.68. *If Ω is a collection of subsets of a nonempty set A (not necessarily a partition) such that*

$$\bigcup_{X \in \Omega} X = A,$$

then R_Ω is reflexive.

Problem 7.69. Is it necessary to require A to be nonempty in Theorem 7.68?

Theorem 7.70. *If Ω is a collection of subsets of a set A (not necessarily a partition), then R_Ω is symmetric.*

Theorem 7.71. *If Ω is a collection of subsets of a set A (not necessarily a partition) such that the elements of Ω are pairwise disjoint, then R_Ω is transitive.*

Problem 7.72. Why didn't we require A to be nonempty in Theorems 7.70 and 7.71?

Recall that Theorem 7.59 says that the equivalence classes for a relation on a nonempty set A determines a partition of A. The following theorem tells us that every partition of a set yields an equivalence relation where the equivalence classes correspond to the blocks of the partition. This result is a consequence of Theorems 7.68, 7.70, and 7.71.

Theorem 7.73. *If Ω is a partition of a set A, then R_Ω is an equivalence relation.*

Together, Theorems 7.59 and 7.73 tell us that equivalence relations and partitions are two different ways of viewing the same thing.

Corollary 7.74. *If R is a relation on a nonempty set A such that the collection of the set of relatives with respect to R is a partition of A, then R is an equivalence relation.*

Problem 7.75. Let $A = \{\circ, \triangle, \blacktriangle, \square, \blacksquare, \bigstar, ☺, ☹\}$. Make up a partition Ω on A and then draw the digraph corresponding to R_Ω.

> In the broad light of day mathematicians check their equations and their proofs, leaving no stone unturned in their search for rigour. But, at night, under the full moon, they dream, they float among the stars and wonder at the miracle of the heavens. They are inspired. Without dreams there is no art, no mathematics, no life.
>
> Michael Atiyah, mathematician

7.4 Modular Arithmetic

In this section, we look at a particular family of equivalence relations on the integers and explore the way in which arithmetic interacts with them.

Definition 7.76. For each $n \in \mathbb{N}$, define $n\mathbb{Z}$ to be the set of all integers that are divisible by n. In set-builder notation, we have

$$\boxed{n\mathbb{Z} := \{m \in \mathbb{Z} \mid m = nk \text{ for some } k \in \mathbb{Z}\}}.$$

For example, $5\mathbb{Z} = \{\dots, -10, -5, 0, 5, 10, \dots\}$ and $2\mathbb{Z}$ is the set of even integers.

Problem 7.77. Consider the sets $3\mathbb{Z}$, $5\mathbb{Z}$, $15\mathbb{Z}$, and $20\mathbb{Z}$.

(a) List at least five elements in each of the above sets.

(b) Notice that $3\mathbb{Z} \cap 5\mathbb{Z} = n\mathbb{Z}$ for some $n \in \mathbb{N}$. What is n? Describe $15\mathbb{Z} \cap 20\mathbb{Z}$ in a similar way.

(c) Draw a Venn diagram illustrating how the sets $3\mathbb{Z}$, $5\mathbb{Z}$, and $15\mathbb{Z}$ intersect.

(d) Draw a Venn diagram illustrating how the sets $5\mathbb{Z}$, $15\mathbb{Z}$, and $20\mathbb{Z}$ intersect.

Theorem 7.78. *Let $n \in \mathbb{N}$. If $a, b \in n\mathbb{Z}$, then $-a$, $a + b$, and ab are also in $n\mathbb{Z}$.*

Definition 7.79. For each $n \in \mathbb{N}$, define the relation $\boxed{\equiv_n}$ on $\mathbb{Z}$ via $a \equiv_n b$ if $a - b \in n\mathbb{Z}$. We read $a \equiv_n b$ as "a is congruent to b modulo n."

Note that $a-b \in n\mathbb{Z}$ if and only if n divides $a-b$, which implies that $a \equiv_n b$ if and only if n divides $a-b$.

Example 7.80. We encountered $\equiv_5$ in Problem 7.22 and discovered that there were five distinct sets of relatives. In particular, we have

$$\begin{aligned}
\operatorname{rel}(0) &= \{\dots,-10,-5,0,5,10,\dots\}\\
\operatorname{rel}(1) &= \{\dots,-9,-4,1,6,11,\dots\}\\
\operatorname{rel}(2) &= \{\dots,-8,-3,2,7,12,\dots\}\\
\operatorname{rel}(3) &= \{\dots,-7,-2,3,8,13,\dots\}\\
\operatorname{rel}(4) &= \{\dots,-6,-1,4,9,14,\dots\}.
\end{aligned}$$

Notice that this collection forms a partition of $\mathbb{Z}$. By Corollary 7.74, the relation $\equiv_5$ must be an equivalence relation.

The previous example generalizes as expected. You can prove the following theorem by either proving that $\equiv_n$ is reflexive, symmetric, and transitive or by utilizing Corollary 7.74.

Theorem 7.81. *For $n \in \mathbb{N}$, the relation $\equiv_n$ is an equivalence relation on $\mathbb{Z}$.*

We have have special notation and terminology for the equivalence classes that correspond to $\equiv_n$.

Definition 7.82. For $n \in \mathbb{N}$, let $[a]_n$ denote the equivalence class of a with respect to $\equiv_n$ (see Definitions 7.17 and 7.44). The equivalence class $[a]_n$ is called the **congruence** (or **residue**) **class of** a **modulo** n. The collection of all equivalence classes determined by $\equiv_n$ is denoted $\mathbb{Z}/n\mathbb{Z}$, which is read "$\mathbb{Z}$ modulo $n\mathbb{Z}$".

Example 7.83. Let's compute $[2]_7$. Tracing back through the definitions, we see that

$$\begin{aligned}
m \in [2]_7 &\iff m \equiv_7 2\\
&\iff m-2 \in 7\mathbb{Z}\\
&\iff m-2 = 7k \text{ for some } k \in \mathbb{Z}\\
&\iff m = 7k+2 \text{ for some } k \in \mathbb{Z}.
\end{aligned}$$

Since the multiples of 7 are $7\mathbb{Z} = \{\dots,-14,-7,0,7,14,\dots\}$, we can find $[2]_7$ by adding 2 to each element of $7\mathbb{Z}$ to get $[2]_7 = \{\dots,-12,-5,2,9,16,\dots\}$.

Problem 7.84. For each of the following congruence classes, find five elements in the set such that at least one is greater than 70 and one is less than 70.

(a) $[4]_7$

(b) $[-3]_7$

(c) $[7]_7$

Problem 7.85. Describe $[0]_3$, $[1]_3$, $[2]_3$, $[4]_3$, and $[-2]_3$ as lists of elements as in Example 7.83. How many distinct congruence classes are in $\mathbb{Z}/3\mathbb{Z}$? Theorem 7.43 might be helpful.

Consider using Theorem 7.42 to prove the next theorem.

Theorem 7.86. *For $n \in \mathbb{N}$ and $a, b \in \mathbb{Z}$, $[a]_n = [b]_n$ if and only if n divides $a - b$.*

Corollary 7.87. *For $n \in \mathbb{N}$ and $a \in \mathbb{Z}$, $[a]_n = [0]_n$ if and only if n divides a.*

The next result provides a useful characterization for when two congruence classes are equal. The Division Algorithm will be useful when proving this theorem.

Theorem 7.88. *For $n \in \mathbb{N}$ and $a, b \in \mathbb{Z}$, $[a]_n = [b]_n$ if and only if a and b have the same remainder when divided by n.*

When proving Part (a) of the next theorem, make use of Theorem 7.86. For Part (b), note that $a_1b_1 - a_2b_2 = a_1b_1 - a_2b_1 + a_2b_1 - a_2b_2$.

Theorem 7.89. *Let $n \in \mathbb{N}$ and let $a_1, a_2, b_1, b_2 \in \mathbb{Z}$. If $[a_1]_n = [a_2]_n$ and $[b_1]_n = [b_2]_n$, then*

(a) $[a_1 + b_1]_n = [a_2 + b_2]_n$, *and*

(b) $[a_1 \cdot b_1]_n = [a_2 \cdot b_2]_n$.

The previous theorem allows us to define addition and multiplication in $\mathbb{Z}/n\mathbb{Z}$.

Definition 7.90. Let $n \in \mathbb{N}$. We define the sum and product of congruence classes in $\mathbb{Z}/n\mathbb{Z}$ via

$$[a]_n + [b]_n := [a + b]_n \quad \text{and} \quad [a]_n \cdot [b]_n := [a \cdot b]_n.$$

Example 7.91. By Definition 7.90, $[2]_7 + [6]_7 = [2+6]_7 = [8]_7$. By Theorem 7.86, $[8]_7 = [1]_7$, and so $[2]_7 + [6]_7 = [1]_7$. Similarly, $[2]_7 \cdot [6]_7 = [2 \cdot 6]_7 = [12]_7 = [5]_7$.

Addition and multiplication for $\mathbb{Z}/n\mathbb{Z}$ has many familiar—and some not so familiar—properties. For example, addition and multiplication of congruence classes are both associative and commutative. However, it is possible for $[a]_n \cdot [b]_n = [0]_n$ even when $[a]_n \neq [0]_n$ and $[b]_n \neq [0]_n$.

Theorem 7.92. *If $n \in \mathbb{N}$, then addition in $\mathbb{Z}/n\mathbb{Z}$ is commutative and associative. That is, for all $[a]_n, [b]_n, [c]_n \in \mathbb{Z}/n\mathbb{Z}$, we have*

(a) $[a]_n + [b]_n = [b]_n + [a]_n$, *and*

(b) $([a]_n + [b]_n) + [c]_n = [a]_n + ([b]_n + [c]_n)$.

Theorem 7.93. *If $n \in \mathbb{N}$, then multiplication in $\mathbb{Z}/n\mathbb{Z}$ is commutative and associative. That is, for all $[a]_n, [b]_n, [c]_n \in \mathbb{Z}/n\mathbb{Z}$, we have*

(a) $[a]_n \cdot [b]_n = [b]_n \cdot [a]_n$, *and*

(b) $([a]_n \cdot [b]_n) \cdot [c]_n = [a]_n \cdot ([b]_n \cdot [c]_n)$.

One consequence of Theorems 7.92(b) and 7.93(b) is that parentheses are not needed when adding or multiplying congruence classes. The next theorem follows from Definition 7.90 together with Theorems 7.92(b) and 7.93(b) and induction on k.

Theorem 7.94. *Let $n \in \mathbb{N}$. For all $k \in \mathbb{N}$, if $[a_1]_n, [a_2]_n, \ldots, [a_k]_n \in \mathbb{Z}/n\mathbb{Z}$, then*

(a) $[a_1]_n + [a_2]_n + \cdots + [a_k]_n = [a_1 + a_2 + \cdots + a_k]_n$, *and*

(b) $[a_1]_n[a_2]_n \cdots [a_k]_n = [a_1 a_2 \cdots a_k]_n$.

The next result is a special case of Theorem 7.94(b).

Corollary 7.95. *Let $n \in \mathbb{N}$. If $a \in \mathbb{Z}$ and $k \in \mathbb{N}$, then $([a]_n)^k = [a^k]_n$.*

Example 7.96. Let's compute $[8^{179}]_7$. We see that

$$\begin{aligned}
[8^{179}]_7 &= ([8]_7)^{179} && \text{(Corollary 7.95)}\\
&= ([1]_7)^{179} && \text{(Theorem 7.86)}\\
&= [1^{179}]_7 && \text{(Corollary 7.95)}\\
&= [1]_7.
\end{aligned}$$

For Part (a) in the next problem, use the fact that $[6]_7 = [-1]_7$. For Part (b), use the fact that $[2^3]_7 = [1]_7$.

Problem 7.97. For each of the following, find a number a with $0 \leq a \leq 6$ such that the given quantity is equal to $[a]_7$.

(a) $[6^{179}]_7$

(b) $[2^{300}]_7$

(c) $[2^{301} + 5]_7$

Problem 7.98. Find a and b such that $[a]_6 \cdot [b]_6 = [0]_6$ but $[a]_6 \neq [0]_6$ and $[b]_6 \neq [0]_6$.

Theorem 7.99. *If $n \in \mathbb{N}$ such that n is not prime, then there exists $[a]_n, [b]_n \in \mathbb{Z}/n\mathbb{Z}$ such that $[a]_n \cdot [b]_n = [0]_n$ while $[a]_n \neq [0]_n$ and $[b]_n \neq [0]_n$.*

Problem 7.100. Notice that $2x = 1$ has no solution in $\mathbb{Z}$. Show that $[2]_7[x]_7 = [1]_7$ does have a solution with x in $\mathbb{Z}$. What about $[14]_7[x]_7 = [1]_7$?

Make use of Theorem 7.94, Corollary 7.95, and Theorem 7.86 to prove the following theorem.

Theorem 7.101. *If $m \in \mathbb{N}$ such that*

$$m = a_k 10^k + a_{k-1} 10^{k-1} + \cdots + a_1 10 + a_0,$$

where $a_k, a_{k-1}, \ldots, a_1, a_0 \in \{0, 1, \ldots, 9\}$ (i.e., $a_k, a_{k-1}, \ldots, a_1, a_0$ are the digits of m), then

$$[m]_3 = [a_k + a_{k-1} + \cdots + a_1 + a_0]_3.$$

You likely recognize the next result. Its proof follows quickly from Corollary 7.87 together with the previous theorem.

Theorem 7.102. *An integer is divisible by 3 if and only if the sum of its digits is divisible by 3.*

Let's revisit Theorem 4.21, which we originally proved by induction.

Problem 7.103. Use Corollary 7.87 to prove that for all integers $n \geq 0$, $3^{2n} - 1$ is divisible by 8. You will need to handle the case involving $n = 0$ separately.

We close this chapter with a fun problem.

Problem 7.104. Prove or provide a counterexample: No integer n exists such that $4n + 3$ is a perfect square.

> Without change something sleeps inside us, and seldom awakens. The sleeper must awaken.
>
> ---
>
> *Dune* by Frank Herbert

I write one page of masterpiece to ninety-one pages of shit.

Ernest Hemingway, novelist & journalist

8 Functions

In this chapter, we will introduce the concept of function as a special type of relation. Our definition should agree with any previous definition of function that you may have learned. We will also study various properties that a function may or may not possess.

8.1 Introduction to Functions

Up until this point, you may have only encountered functions as an algebraic rule, e.g., $f(x) = x^2 - 1$, for transforming one real number into another. However, we can study functions in a much broader context. The basic building blocks of a function are a first set and a second set, say X and Y, and a "correspondence" that assigns *every* element of X to *exactly one* element of Y. Let's take a look at the actual definition.

Definition 8.1. Let X and Y be two nonempty sets. A **function f from X to Y** is a relation from X to Y such that for every $x \in X$, there exists a unique $y \in Y$ such that $(x, y) \in f$. The set X is called the **domain** of f and is denoted by $\boxed{\operatorname{Dom}(f)}$. The set Y is called the **codomain** of f and is denoted by $\boxed{\operatorname{Codom}(f)}$ while the subset of the codomain defined via

$$\boxed{\operatorname{Rng}(f) := \{y \in Y \mid \text{there exists } x \text{ such that } (x, y) \in f\}}$$

is called the **range** of f or the **image of** X under f.

There is a variety of notation and terminology associated to functions. We will write $\boxed{f : X \to Y}$ to indicate that f is a function from X to Y. We will make use of statements such as "Let $f : X \to Y$ be the function defined via..." or "Define $f : X \to Y$ via...", where f is understood to be a function in the second statement. Sometimes the word **mapping** (or **map**) is used in place of the word function. If $(a, b) \in f$ for a function f, we often write $\boxed{f(a) = b}$ and say that "f maps a to b" or "f of a equals b". In this case, a may be called an **input** of f and is the **preimage** of b under f while b is called an **output** of f and is the **image** of a under f. Note that the domain of a function is the set of inputs while the range is the set of outputs for the function.

According to our definition, if $f : X \to Y$ is a function, then every element of the domain is utilized exactly once. However, there are no restrictions on whether an element of the codomain ever appears in the second coordinate of an ordered pair in the relation. Yet if an element of Y is in the range of f, it may appear in more than one ordered pair in the relation.

It follows immediately from the definition of function that two functions are equal if and only if they have the same domain, same codomain, and the same set of ordered pairs in the relation. That is, functions f and g are equal if and only if $\text{Dom}(f) = \text{Dom}(g)$, $\text{Codom}(f) = \text{Codom}(g)$, and $f(x) = g(x)$ for all $x \in X$.

Since functions are special types of relations, we can represent them using digraphs and graphs when practical. Digraphs for functions are often called **function** (or **mapping**) **diagrams**. When drawing function diagrams, it is standard practice to put the vertices for the domain on the left and the vertices for the codomain on the right, so that all directed edges point from left to right. We may also draw an additional arrow labeled by the name of the function from the domain to the codomain.

Example 8.2. Let $X = \{a, b, c, d\}$ to $Y = \{1, 2, 3, 4\}$ and define the relation f from X to Y via

$$f = \{(a, 2), (b, 4), (c, 4), (d, 1)\}.$$

Since each element X appears exactly once as a first coordinate, f is a function with domain X and codomain Y (i.e., $f : X \to Y$). In this case, we see that $\text{Rng}(f) = \{1, 2, 4\}$. Moreover, we can write things like $f(a) = 2$ and $c \mapsto 4$, and say things like "f maps b to 4" and "the image of d is 1." The function diagram for f is depicted in Figure 8.1.

Problem 8.3. Determine whether each of the relations defined in the following examples and problems is a function.

(a) Example 7.3 (see Figure 7.1)

(b) Example 7.14 (see Figure 7.3)

(c) Problem 7.15

(d) Problem 7.21

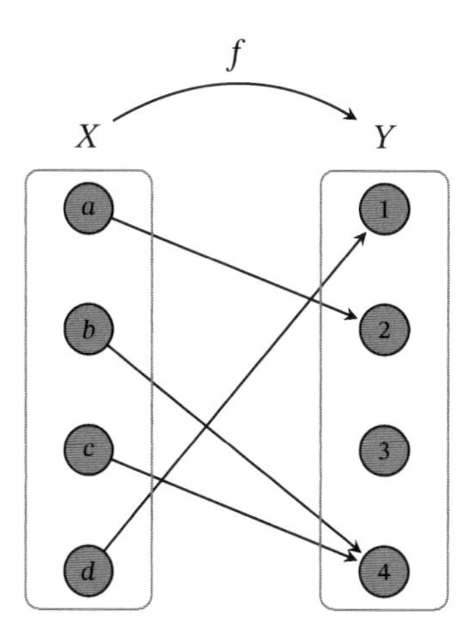

Figure 8.1. Function diagram for a function from $X = \{a, b, c, d, \}$ to $Y = \{1, 2, 3, 4\}$.

Problem 8.4. Let $X = \{\circ, \square, \triangle, \smiley\}$ and $Y = \{a, b, c, d, e\}$. For each of the following relations, draw the corresponding digraph and determine whether the relation represents a function from X to Y, Y to X, X to X, or does not represent a function. If the relation is a function, determine the domain, codomain, and range.

(a) $f = \{(\circ, a), (\square, b), (\triangle, c), (\smiley, d)\}$

(b) $g = \{(\circ, a), (\square, b), (\triangle, c), (\smiley, c)\}$

(c) $h = \{(\circ, a), (\square, b), (\triangle, c), (\circ, d)\}$

(d) $k = \{(\circ, a), (\square, b), (\triangle, c), (\smiley, d), (\square, e)\}$

(e) $l = \{(\circ, e), (\square, e), (\triangle, e), (\smiley, e)\}$

(f) $m = \{(\circ, a), (\triangle, b), (\smiley, c)\}$

(g) $i = \{(\circ, \circ), (\square, \square), (\triangle, \triangle), (\smiley, \smiley)\}$

(h) Define the relation happy from Y to X via $(y, \smiley) \in$ happy for all $y \in Y$.

(i) nugget $= \{(\circ, \circ), (\square, \square), (\triangle, \triangle), (\smiley, \square)\}$

The last two parts of the previous problem make it clear that functions may have names consisting of more than one letter. The function names sin, cos, log, and ln are instances of this that you have likely encountered in your previous experience in mathematics. One thing that you may have never noticed is the type of font that we use for function names. It is common to italicize generic function names like f but not common function names like sin. However, we always italicize the variables used to represent the input and output for a function. For example, consider the font types used in the expressions $\sin(x)$ and $\ln(a)$.

Problem 8.5. What properties does the digraph for a relation from X to Y need to have in order for it to represent a function?

Problem 8.6. In high school you may have been told that a graph represents a function if it passes the **vertical line test**. Carefully state what the vertical line test says and then explain why it works.

Sometimes we can define a function using a formula. For example, we can write $f(x) = x^2 - 1$ to mean that each x in the domain of f maps to $x^2 - 1$ in the codomain. However, notice that providing only a formula is ambiguous! A function is determined by its domain, codomain, and the correspondence between these two sets. If we only provide a description for the correspondence, it is not clear what the domain and codomain are. Two functions that are defined by the same formula, but have different domains or codomains are *not* equal.

Example 8.7. The function $f : \mathbb{R} \to \mathbb{R}$ defined via $f(x) = x^2 - 1$ is not equal to the function $g : \mathbb{N} \to \mathbb{R}$ defined by $g(x) = x^2 - 1$ since the two functions do not have the same domain.

Sometimes we rely on context to interpret the domain and codomain. For example, in a calculus class, when we describe a function in terms of a formula, we are implicitly assuming that the domain is the largest allowable subset of $\mathbb{R}$—sometimes called the **default domain**—that makes sense for the given formula while the codomain is $\mathbb{R}$.

Example 8.8. If we write $f(x) = x^2 - 1$, $g(x) = \sqrt{x}$, and $h(x) = \frac{1}{x}$ without mentioning the domains, we would typically interpret these as the functions $f : \mathbb{R} \to \mathbb{R}$, $g : [0, \infty) \to \mathbb{R}$, and $h : \mathbb{R} \setminus \{0\} \to \mathbb{R}$ that are determined by their respective formulas.

Problem 8.9. Provide an example of each of the following. You may draw a function diagram, write down a list of ordered pairs, or write a formula as long as the domain and codomain are clear.

(a) A function f from a set with 4 elements to a set with 3 elements such that $\operatorname{Rng}(f) = \operatorname{Codom}(f)$.

(b) A function g from a set with 4 elements to a set with 3 elements such that $\operatorname{Rng}(g)$ is strictly smaller than $\operatorname{Codom}(g)$.

Problem 8.10. Let $f : X \to Y$ be a function and suppose that X and Y are finite sets with n and m elements, respectively, such that $n < m$. Is it possible for $\text{Rng}(f) = \text{Codom}(f)$? If so, provide an example. If this is not possible, explain why.

There are a few special functions that we should know the names of.

Definition 8.11. If X and Y are nonempty sets such that $X \subseteq Y$, then the function $\iota : X \to Y$ defined via $\iota(x) = x$ is called the **inclusion map from X into Y**.

Note that "ι" is the Greek letter "iota".

Problem 8.12. Let $X = \{a, b, c\}$ and $Y = \{a, b, c, d\}$. Draw the function diagram of the inclusion map from X into Y.

If the domain and codomain are equal, the inclusion map has a special name.

Definition 8.13. If X is a nonempty set, then the function $i_X : X \to X$ defined via $i_X(x) = x$ is called the **identity map** (or **identity function**) **on** X.

Example 8.14. The relation defined in Problem 8.4(g) is the identity map on $X = \{\circ, \square, \triangle, ☺\}$.

Problem 8.15. Draw a portion of the graph of the identity map on $\mathbb{R}$ as a subset of $\mathbb{R}^2$.

Problem 8.16. Let A be a nonempty set.

(a) Suppose R is an equivalence relation on A. What conditions on R must hold in order for R to be a function from A to A?

(b) Suppose $f : A \to A$ is a function. Under what conditions is f an equivalence relation on A?

Definition 8.17. Any function $f : X \to Y$ defined via $f(x) = c$ for a fixed $c \in Y$ is called a **constant function**.

Example 8.18. The function defined in Problem 8.4(h) is an example of a constant function. Notice that we can succinctly describe this function using the formula $\text{happy}(y) = ☺$.

Definition 8.19. A **piecewise-defined function** (or **piecewise function**) is a function defined by specifying its output on a partition of the domain.

Note that "piecewise" is a way of expressing the function, rather than a property of the function itself.

Example 8.20. We can express the function in Problem 8.4(i) as a piecewise function using the formula

$$\text{nugget}(x) = \begin{cases} x, & \text{if } x \text{ is a geometric shape,} \\ \square, & \text{otherwise.} \end{cases}$$

Example 8.21. The function $f : \mathbb{R} \to \mathbb{R}$ defined via

$$f(x) = \begin{cases} x^2 - 1, & \text{if } x \geq 0, \\ 17, & \text{if } -2 \leq x < 0, \\ -x, & \text{if } x < -2 \end{cases}$$

is an example of a piecewise-defined function.

Problem 8.22. Define $f : \mathbb{R} \setminus \{0\} \to \mathbb{R}$ via $f(x) = \frac{|x|}{x}$. Express f as a piecewise function.

It is important to point out that not every function can be described using a formula! Despite your prior experience, functions that can be represented succinctly using a formula are rare.

The next problem illustrates that some care must be taken when attempting to define a function.

Problem 8.23. For each of the following, explain why the given description does not define a function.

(a) Define $f : \{1, 2, 3\} \to \{1, 2, 3\}$ via $f(a) = a - 1$.

(b) Define $g : \mathbb{N} \to \mathbb{Q}$ via $g(n) = \frac{n}{n-1}$.

(c) Let $A_1 = \{1, 2, 3\}$ and $A_2 = \{3, 4, 5\}$. Define $h : A_1 \cup A_2 \to \{1, 2\}$ via

$$h(x) = \begin{cases} 1, & \text{if } x \in A_1 \\ 2, & \text{if } x \in A_2. \end{cases}$$

(d) Define $s : \mathbb{Q} \to \mathbb{Z}$ via $s(a/b) = a + b$.

In mathematics, we say that an expression is **well defined** (or **unambiguous**) if its definition yields a unique interpretation. Otherwise, we say that the expression is not well defined (or is **ambiguous**). For example, if $a, b, c \in \mathbb{R}$, then the expression abc is well defined since it does not matter if we interpret this as $(ab)c$ or $a(bc)$ since the real numbers are associative under multiplication. This issue was lurking behind the scenes in the statement of Theorem 7.94. In particular, the expressions

$$[a_1]_n + [a_2]_n + \cdots + [a_k]_n$$

and

$$[a_1]_n[a_2]_n \cdots [a_k]_n$$

are well defined in $\mathbb{Z}/n\mathbb{Z}$ in light of Theorems 7.92(b) and 7.93(b).

When we attempt to define a function, it may not be clear without doing some work that our definition really does yield a function. If there is some potential ambiguity in the definition of a function that ends up not causing any issues, we say that the function is well defined. However, this phrase is a bit of misnomer since all functions are well defined. The issue of whether a description for a proposed function is well defined often arises when defining things in terms of representatives of equivalence classes, or more generally in terms of how an element of the domain is written. For example, the descriptions given in Parts (c) and (d) of Problem 8.23 are not well defined. To show that a potentially ambiguous description for a function $f : X \to Y$ is well defined prove that if a and b are two representations for the same element in X, then $f(a) = f(b)$.

Problem 8.24. For each of the following, determine whether the description determines a well-defined function.

(a) Define $f : \mathbb{Z}/5\mathbb{Z} \to \mathbb{N}$ via

$$f([a]_5) = \begin{cases} 0, & \text{if } a \text{ is even} \\ 1, & \text{if } a \text{ is odd.} \end{cases}$$

(b) Define $g : \mathbb{Z}/6\mathbb{Z} \to \mathbb{N}$ via

$$g([a]_6) = \begin{cases} 0, & \text{if } a \text{ is even} \\ 1, & \text{if } a \text{ is odd.} \end{cases}$$

(c) Define $m : \mathbb{Z}/8\mathbb{Z} \to \mathbb{Z}/10\mathbb{Z}$ via $m([x]_8) = [6x]_{10}$.

(d) Define $h : \mathbb{Z}/10\mathbb{Z} \to \mathbb{Z}/10\mathbb{Z}$ via $h([x]_{10}) = [6x]_{10}$.

(e) Define $k : \mathbb{Z}/43\mathbb{Z} \to \mathbb{Z}/43\mathbb{Z}$ via $k([x]_{43}) = [11x - 5]_{43}$.

(f) Define $\ell : \mathbb{Z}/15\mathbb{Z} \to \mathbb{Z}/15\mathbb{Z}$ via $\ell([x]_{15}) = [5x - 11]_{15}$.

Problem 8.25. Let $k, n \in \mathbb{N}$ and $m \in \mathbb{Z}$. Under what conditions will $f_m : \mathbb{Z}/n\mathbb{Z} \to \mathbb{Z}/k\mathbb{Z}$ given by $f_m([x]_n) = [mx]_k$ be a well-defined function? Prove your claim.

> Don't let anyone rob you of your imagination, your creativity, or your curiosity. It's your place in the world; it's your life. Go on and do all you can with it, and make it the life you want to live.
>
> Mae Jemison, NASA astronaut

8.2 Injective and Surjective Functions

We now turn our attention to some important properties that a function may or may not possess. Recall that if f is a function, then every element in its domain is mapped to a unique element in the range. However, there are no restrictions on whether more than one element of the domain is mapped to the same element in the range. If each element in the range has a unique element in the domain mapping to it, then we say that the function is injective. Moreover, the range of a function is not required to be all of the codomain. If every element of the codomain has at least one element in the domain that maps to it, then we say that the function is surjective. Let's make these definitions a bit more precise.

Definition 8.26. Let $f : X \to Y$ be a function.

(a) The function f is said to be **injective** (or **one-to-one**) if for all $y \in \mathrm{Rng}(f)$, there is a unique $x \in X$ such that $y = f(x)$.

(b) The function f is said to be **surjective** (or **onto**) if for all $y \in Y$, there exists $x \in X$ such that $y = f(x)$.

(c) If f is both injective and surjective, we say that f is **bijective**.

Problem 8.27. Compare and contrast the following statements. Do they mean the same thing?

(a) For all $x \in X$, there exists a unique $y \in Y$ such that $f(x) = y$.

(b) For all $y \in \mathrm{Rng}(f)$, there is a unique $x \in X$ such that $y = f(x)$.

Problem 8.28. Assume that X and Y are finite sets. Provide an example of each of the following. You may draw a function diagram, write down a list of ordered pairs, or write a formula as long as the domain and codomain are clear.

(a) A function $f : X \to Y$ that is injective but not surjective.

(b) A function $f : X \to Y$ that is surjective but not injective.

(c) A function $f : X \to Y$ that is a bijection.

(d) A function $f : X \to Y$ that is neither injective nor surjective.

Problem 8.29. Provide an example of each of the following. You may either draw a graph or write down a formula. Make sure you have the correct domain.

(a) A function $f : \mathbb{R} \to \mathbb{R}$ that is injective but not surjective.

(b) A function $f : \mathbb{R} \to \mathbb{R}$ that is surjective but not injective.

(c) A function $f : \mathbb{R} \to \mathbb{R}$ that is a bijection.

(d) A function $f : \mathbb{R} \to \mathbb{R}$ that is neither injective nor surjective.

(e) A function $f : \mathbb{N} \times \mathbb{N} \to \mathbb{N}$ that is injective.

Problem 8.30. Suppose $X \subseteq \mathbb{R}$ and $f : X \to \mathbb{R}$ is a function. Fill in the blank with the appropriate word.

> The function $f : X \to \mathbb{R}$ is ________ if and only if every horizontal line hits the graph of f *at most once.*

This statement is often called the **horizontal line test**. Explain why the horizontal line test is true.

Problem 8.31. Suppose $X \subseteq \mathbb{R}$ and $f : X \to \mathbb{R}$ is a function. Fill in the blank with the appropriate word.

> The function $f : X \to \mathbb{R}$ is ________ if and only if every horizontal line hits the graph of f *at least once.*

Explain why this statement is true.

Problem 8.32. Suppose $X \subseteq \mathbb{R}$ and $f : X \to \mathbb{R}$ is a function. Fill in the blank with the appropriate word.

> The function $f : X \to \mathbb{R}$ is ________ if and only if every horizontal line hits the graph of f *exactly once.*

Explain why this statement is true.

How do we prove that a function f is injective? We would need to show that every element in the range has a unique element from the domain that maps to it. First, notice that each element in the range can be written as $f(x)$ for at least one x in the domain. To argue that each such element in domain is unique, we can suppose $f(x_1) = f(x_2)$ for arbitrary x_1 and x_2 in the domain and then work to show that $x_1 = x_2$. It is important to point out that when we suppose $f(x_1) = f(x_2)$ for some x_1 and x_2, we are not assuming that x_1 and x_2 are different. In general, when we write "Let $x_1, x_2 \in X\ldots$", we are leaving open the possibility that x_1 and x_2 are actually the same element. One could approach proving that a function is injective by utilizing a proof by contradiction, but this is not usually necessary.

Skeleton Proof 8.33 (Proof that a function is injective). Here is the general structure for proving that a function is injective.

> *Proof.* Assume $f : X \to Y$ is a function defined by (or satisfying)... [*Use the given definition (or describe the given property) of f*]. Let $x_1, x_2 \in X$ and suppose $f(x_1) = f(x_2)$.
>
> ... [*Use the definition (or property) of f to verify that $x_1 = x_2$*] ...
>
> Therefore, the function f is injective. □

How do we prove that a function f is surjective? We would need to argue that every element in the codomain is also in the range. Sometimes, the proof that a particular function is surjective is extremely short, so do not second guess yourself if you find yourself in this situation.

Skeleton Proof 8.34 (Proof that a function is surjective). Here is the general structure for proving that a function is surjective.

> *Proof.* Assume $f : X \to Y$ is a function defined by (or satisfying)... [*Use the given definition (or describe the given property) of f*]. Let $y \in Y$.
>
> ... [*Use the definition (or property) of f to find some $x \in X$ such that $f(x) = y$*] ...
>
> Therefore, the function f is surjective. □

Problem 8.35. Determine whether each of the following functions is injective, surjective, both, or neither. In each case, you should provide a proof or a counterexample as appropriate.

(a) Define $f : \mathbb{R} \to \mathbb{R}$ via $f(x) = x^2$

(b) Define $g : \mathbb{R} \to [0, \infty)$ via $g(x) = x^2$

(c) Define $h : \mathbb{R} \to \mathbb{R}$ via $h(x) = x^3$

(d) Define $k : \mathbb{R} \to \mathbb{R}$ via $k(x) = x^3 - x$

(e) Define $c : \mathbb{R} \times \mathbb{R} \to \mathbb{R}$ via $c(x, y) = x^2 + y^2$

(f) Define $f : \mathbb{N} \to \mathbb{N} \times \mathbb{N}$ via $f(n) = (n, n)$

(g) Define $g : \mathbb{Z} \to \mathbb{Z}$ via

$$g(n) = \begin{cases} \frac{n}{2}, & \text{if } n \text{ is even} \\ \frac{n+1}{2}, & \text{if } n \text{ is odd} \end{cases}$$

(h) Define $\ell : \mathbb{Z} \to \mathbb{N}$ via

$$\ell(n) = \begin{cases} 2n+1, & \text{if } n \geq 0 \\ -2n, & \text{if } n < 0 \end{cases}$$

(i) The function h defined in Problem 8.24(d)

(j) The function k defined in Problem 8.24(e)

(k) The function ℓ defined in Problem 8.24(f)

Problem 8.36. Suppose X and Y are nonempty sets with m and n elements, respectively, where $m \leq n$. How many injections are there from X to Y?

Problem 8.37. Compare and contrast the definition of "function" with the definition of "injective function". Consider the vertical line test and horizontal line test in your discussion. Moreover, attempt to capture what it means for a relation to not be a function and for a function to not be an injection by drawing portions of a digraph.

The next two theorems should not come as as surprise.

Theorem 8.38. *The inclusion map $\iota : X \to Y$ for $X \subseteq Y$ is an injection.*

Theorem 8.39. *The identity function $i_X : X \to X$ is a bijection.*

Problem 8.40. Let A and B be nonempty sets and let S be a nonempty subset of $A \times B$. Define $\pi_1 : S \to A$ and $\pi_2 : S \to B$ via $\pi_1(a,b) = a$ and $\pi_2(a,b) = b$. We call π_1 and π_2 the **projections** of S onto A and B, respectively.

(a) Provide examples to show that π_1 does not need to be injective nor surjective.

(b) Suppose that S is also a function. Is π_1 injective? Is π_1 surjective? How about π_2?

The next theorem says that if we have an equivalence relation on a nonempty set, the mapping that assigns each element to its respective equivalence class is a surjective function.

Theorem 8.41. *If $\sim$ is an equivalence relation on a nonempty set A, then the function $f : A \to A/\sim$ defined via $f(x) = [x]$ is a surjection.*

The function from the previous theorem is sometimes called the **canonical projection map** induced by $\sim$.

Problem 8.42. Under what circumstances would the function from the previous theorem also be injective?

Let's explore whether we can weaken the hypotheses of Theorem 8.41.

Problem 8.43. Let R be a relation on a nonempty set A.

(a) What conditions on R must hold in order for $f : A \to \text{Rel}(R)$ defined via $f(a) = \text{rel}(a)$ to be a function?

(b) What additional conditions, if any, must hold on R in order for f to be a surjective function?

Given any function, we can define an equivalence relation on its domain, where the equivalence classes correspond to the elements that map to the same element of the range.

Theorem 8.44. *Let $f : X \to Y$ be a function and define $\sim$ on X via $a \sim b$ if $f(a) = f(b)$. Then $\sim$ is an equivalence relation on X.*

It follows immediately from Theorem 7.59 that the equivalence classes induced by the equivalence relation in Theorem 8.44 partition the domain of a function.

Problem 8.45. For each of the following, identify the equivalence classes induced by the relation from Theorem 8.44 for the given function.

(a) The function f defined in Example 8.2.

(b) The function c defined in Problem 8.35(e). Can you describe the equivalence classes geometrically?

If f is a function, the equivalence relation in Theorem 8.44 allows us to construct a bijective function whose domain is the set of equivalence classes and whose codomain coincides with the range of f. This is an important idea that manifests itself in many areas of mathematics. One such instance is the First Isomorphism Theorem for Groups, which is a fundamental theorem in a branch of mathematics called group theory. When proving the following theorem, the first thing you should do is verify that the description for $\overline{f}$ is well defined.

Theorem 8.46. *Let $f : X \to Y$ be a function and define $\sim$ on X as in Theorem 8.44. Then the function $\overline{f} : X/\!\sim\, \to \text{Rng}(f)$ defined via $\overline{f}([a]) = f(a)$ is a bijection.*

Here is an analogy for helping understand the content of Theorem 8.46. Suppose we have a collection airplanes filled with passengers and a collection of potential destination cities such that at most one airplane may land at each city. The function f indicates which city each passenger lands at while the function $\overline{f}$ indicates which city each airplane lands at. Moreover, the codomain for the function $\overline{f}$ consists only of the cities that an airplane lands at.

Example 8.47. Let $X = \{a, b, c, d, e, f\}$ and $Y = \{1, 2, 3, 4, 5\}$ and define $\varphi : X \to Y$ via

$$\varphi = \{(a, 1), (b, 1), (c, 2), (d, 4), (e, 4), (f, 4)\}.$$

The function diagram for φ is given in Figure 8.2(a), where we have highlighted the elements of the domain that map to the same element in the range by enclosing them in additional boxes. We see that $\text{Rng}(\varphi) = \{1, 2, 4\}$. The function diagram for the induced map $\overline{\varphi}$ that is depicted in Figure 8.2(b) makes it clear that $\overline{\varphi}$ is a bijection. Note that since $\varphi(a) = \varphi(b)$ and $\varphi(d) = \varphi(e) = \varphi(f)$, it must be the case that $[a] = [b]$ and $[d] = [e] = [f]$ according to Theorem 7.42. Thus, the vertices labeled as $[a]$ and $[d]$ in Figure 8.2(b) could have also been labeled as $[b]$ and $[c]$ or $[d]$, respectively. In terms of our passengers and airplanes analogy, $X = \{a, b, c, d, e, f\}$ is the set of passengers, $Y = \{1, 2, 3, 4, 5\}$ is the set of potential destination cities, $X/\sim = \{[a], [c], [d]\}$ is the set of airplanes, and $\text{Rng}(\varphi) = \{1, 2, 4\}$ is the set of cities that airplanes land at. The equivalence class $[a]$ is the airplane containing the passenger a, and since a and b are on the same plane, $[b]$ is also the plane containing the passenger a.

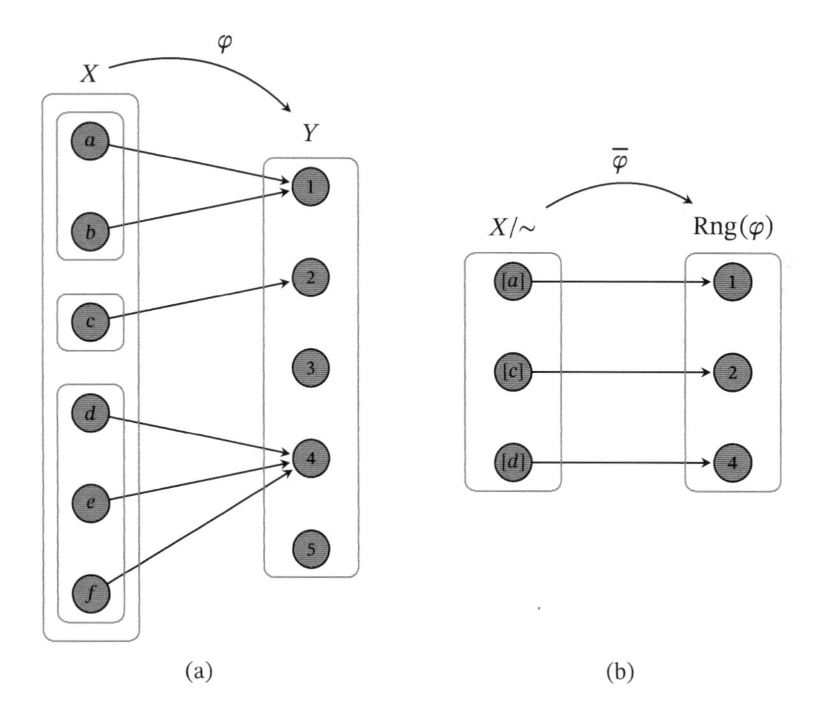

Figure 8.2. Example of a visual representation of Theorem 8.46.

Problem 8.48. Consider the equivalence classes you identified in Parts (a) and (b) of Problem 8.45.

(a) Draw the function diagram for the function $\overline{f}$ as defined in Theorem 8.46, where f is the function defined in Example 8.2.

(b) Geometrically describe the function $\overline{c}$ as defined in Theorem 8.46, where c is the function defined in Problem 8.35(e).

While perhaps not surprising, Problem 8.48(b) tells us that there is a one-to-one correspondence between circles centered at the origin and real numbers.

Problem 8.49. Let $Y = \{0, 1, 2, 3\}$ and define the function $f : \mathbb{Z} \to Y$ such that $f(n)$ equals the unique remainder obtained after dividing n by 4. For example, $f(11) = 3$ since $11 = 4 \cdot 2 + 3$ according to the Division Algorithm (Theorem 6.7). This function is sometimes written as $f(n) = n \pmod 4$, where it is understood that we restrict the output to $\{0, 1, 2, 3\}$. It is clear that f is surjective since 0, 1, 2, and 3 are mapped to 0, 1, 2, and 3, respectively. Figure 8.3 depicts a portion of the function diagram for f, where we have drawn the diagram from the top down instead of left to right.

(a) Describe the equivalence classes induced by the relation given in Theorem 8.44.

(b) What familiar set is $\mathbb{Z}/{\sim}$ equal to?

(c) Draw the function diagram for the function $\overline{f}$ as defined in Theorem 8.46.

(d) The function diagram in Figure 8.3 is a bit hard to interpret due to the ordering of the elements in the domain. Can you find a better way to lay out the vertices in the domain that makes the function f easier to interpret?

Problem 8.50. Consider the function h defined in Problem 8.24(d).

(a) Draw the function diagram for h.

(b) Identify the equivalence classes induced by the relation given in Theorem 8.44.

(c) Draw the function diagram for the function $\overline{h}$ as defined in Theorem 8.46.

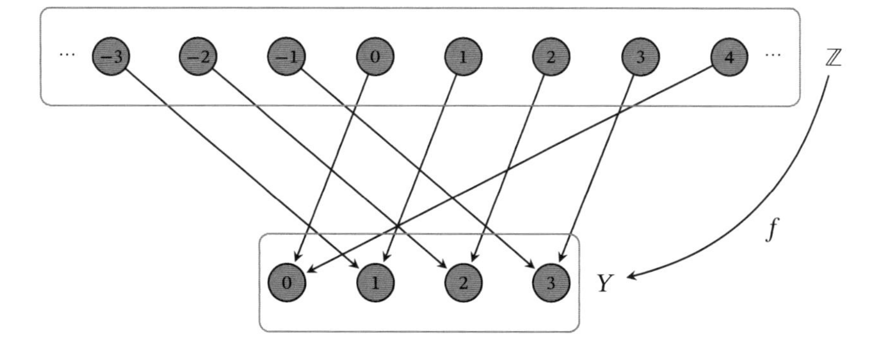

Figure 8.3. Function diagram for the function described in Problem 8.49.

> It is not the critic who counts; not the man who points out how the strong man stumbles, or where the doer of deeds could have done them better. The credit belongs to the man who is actually in the arena, whose face is marred by dust and sweat and blood; who strives valiantly; who errs, who comes short again and again, because there is no effort without error and shortcoming; but who does actually strive to do the deeds; who knows great enthusiasms, the great devotions; who spends himself in a worthy cause; who at the best knows in the end the triumph of high achievement, and who at the worst, if he fails, at least fails while daring greatly, so that his place shall never be with those cold and timid souls who neither know victory nor defeat.
>
> Theodore Roosevelt, statesman & conservationist

8.3 Compositions and Inverse Functions

We begin this section with a method for combining two functions together that have compatible domains and codomains.

Definition 8.51. If $f : X \to Y$ and $g : Y \to Z$ are functions, we define $g \circ f : X \to Z$ via $\boxed{(g \circ f)(x) = g(f(x))}$. The function $g \circ f$ is called the **composition of** f **and** g.

It is important to notice that the function on the right is the one that "goes first." Moreover, we cannot compose any two random functions since the codomain of the first function must agree with the domain of the second function. In particular, $f \circ g$ may not be a sensible function even when $g \circ f$ exists. Figure 8.4 provides a visual representation of function composition in terms of function diagrams.

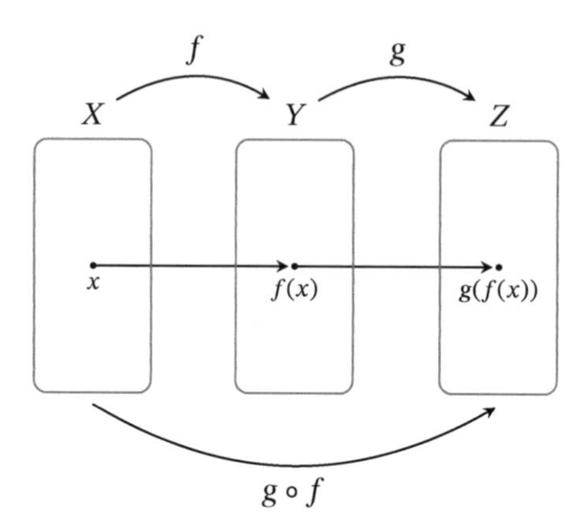

Figure 8.4. Visual representation of function composition.

Problem 8.52. Let $X = \{1, 2, 3, 4\}$ and define $f : X \to X$ and $g : X \to X$ via

$$f = \{(1, 1), (2, 3), (3, 3), (4, 4)\}$$

and

$$g = \{(1, 1), (2, 2), (3, 1), (4, 1)\}.$$

For each of the following functions, draw the corresponding function diagram in the spirit of Figure 8.4 and identify the range.

(a) $g \circ f$

(b) $f \circ g$

The previous problem illustrates that $f \circ g$ and $g \circ f$ need not be equal even when both composite functions exist.

Example 8.53. Consider the inclusion map $\iota : X \to Y$ such that X is a proper subset of Y and suppose $f : Y \to Z$ is a function. Then the composite function $f \circ \iota : X \to Z$ is given by

$$f \circ \iota(x) = f(\iota(x)) = f(x)$$

for all $x \in X$. Notice that $f \circ \iota$ is simply the function f but with a smaller domain. In this case, we say that $f \circ \iota$ is the **restriction of** f **to** X, which is often denoted by $f|_X$.

Problem 8.54. Define $f : \mathbb{R} \to \mathbb{R}$ and $g : \mathbb{R} \to \mathbb{R}$ via $f(x) = x^2$ and $g(x) = 3x - 5$, respectively. Determine formulas for the composite functions $f \circ g$ and $g \circ f$.

Problem 8.55. Define $f : \mathbb{R} \to \mathbb{R}$ and $g : \mathbb{R} \to \mathbb{R}$ via

$$f(x) = \begin{cases} 5x + 7, & \text{if } x < 0 \\ 2x + 1, & \text{if } x \geq 0 \end{cases}$$

and $g(x) = 7x - 11$, respectively. Find a formula for the composite function $g \circ f$.

Problem 8.56. Define $f : \mathbb{Z}/15\mathbb{Z} \to \mathbb{Z}/23\mathbb{Z}$ and $g : \mathbb{Z}/23\mathbb{Z} \to \mathbb{Z}/32\mathbb{Z}$ via $f([x]_{15}) = [3x + 5]_{23}$ and $g([x]_{23}) = [2x + 1]_{32}$, respectively. Find a formula for the composite function $g \circ f$.

The following result provides some insight into where the identity map got its name.

Theorem 8.57. *If $f : X \to Y$ is a function, then $f \circ i_X = f = i_Y \circ f$, where i_X and i_Y are the identity maps on X and Y, respectively.*

The next theorem tells us that function composition is associative.

Theorem 8.58. *If $f : X \to Y$, $g : Y \to Z$, and $h : Z \to W$ are functions, then $(h \circ g) \circ f = h \circ (g \circ f)$.*

Problem 8.59. In each case, give examples of finite sets X, Y, and Z, and functions $f : X \to Y$ and $g : Y \to Z$ that satisfy the given conditions. Drawing a function diagram is sufficient.

(a) f is surjective, but $g \circ f$ is not surjective.

(b) g is surjective, but $g \circ f$ is not surjective.

(c) f is injective, but $g \circ f$ is not injective.

(d) g is injective, but $g \circ f$ is not injective.

Theorem 8.60. *If $f : X \to Y$ and $g : Y \to Z$ are both surjective functions, then $g \circ f$ is also surjective.*

Theorem 8.61. *If $f : X \to Y$ and $g : Y \to Z$ are both injective functions, then $g \circ f$ is also injective.*

Corollary 8.62. *If $f : X \to Y$ and $g : Y \to Z$ are both bijections, then $g \circ f$ is also a bijection.*

Problem 8.63. Assume that $f : X \to Y$ and $g : Y \to Z$ are both functions. Determine whether each of the following statements is true or false. If a statement is true, prove it. Otherwise, provide a counterexample.

(a) If $g \circ f$ is injective, then f is injective.

(b) If $g \circ f$ is injective, then g is injective.

(c) If $g \circ f$ is surjective, then f is surjective.

(d) If $g \circ f$ is surjective, then g is surjective.

Theorem 8.64. *Let $f : X \to Y$ be a function. Then f is injective if and only if there exists a function $g : Y \to X$ such that $g \circ f = i_X$, where i_X is the identity map on X.*

The function g in the previous theorem is often called a **left inverse** of f.

Theorem 8.65. *Let $f : X \to Y$ be a function. Then f is surjective if and only if there exists a function $g : Y \to X$ such that $f \circ g = i_Y$, where i_Y is the identity map on Y.*

The function g in the previous theorem is often called a **right inverse** of f.

Problem 8.66. Let $X = \{a, b\}$ and $Y = \{1, 2\}$.

(a) Provide an example of a function that has a left inverse but does not have a right inverse. Find the left inverse of your proposed function.

(b) Provide an example of a function that has a right inverse but does not have a left inverse. Find the right inverse of your proposed function.

Problem 8.67. Define $f : \mathbb{R} \to \mathbb{R}$ via $f(x) = x^2$. Explain why f does not have a left inverse nor a right inverse.

Problem 8.68. Define $f : \mathbb{R} \to [0, \infty)$ via $f(x) = x^2$ and $g : [0, \infty) \to \mathbb{R}$ via $g(x) = \sqrt{x}$.

(a) Explain why f does not have a left inverse.

(b) Verify that g is the right inverse of f by computing $f \circ g(x)$.

Corollary 8.69. *If $f : X \to Y$ and $g : Y \to X$ are functions satisfying $g \circ f = i_X$ and $f \circ g = i_Y$, then f is a bijection.*

In the previous result, the functions f and g "cancel" each other out. In this case, we say that g is a **two-sided inverse** of f.

Definition 8.70. Let $f : X \to Y$ be a function. The relation f^{-1} from Y to X, called f **inverse**, is defined via

$$\boxed{f^{-1} = \{(f(x), x) \in Y \times X \mid x \in X\}}.$$

Notice that we called f^{-1} a relation and not a function. In some circumstances f^{-1} will be a function and sometimes it will not be. Given a function f, the inverse relation is simply the set of ordered pairs that results from reversing the ordered pairs in f. It is worth pointing out that we have only defined inverse relations for functions. However, one can easily adapt our definition to handle arbitrary relations.

Problem 8.71. Consider the function f given in Example 8.2 (see Figure 8.1). List the ordered pairs in the relation f^{-1} and draw the corresponding digraph. Is f^{-1} a function?

Problem 8.72. Provide an example of a function $f : X \to Y$ such that f^{-1} is a function. Drawing a function diagram is sufficient.

Problem 8.73. Suppose $X \subseteq \mathbb{R}$ and $f : X \to \mathbb{R}$ is a function. What is the relationship between the graph of the function f and the graph of the inverse relation f^{-1}?

Theorem 8.74. *Let $f : X \to Y$ be a function. Then $f^{-1} : Y \to X$ is a function if and only if f is a bijection.*

Problem 8.75. Suppose $f : \mathbb{R} \to \mathbb{R}$ is a function. Fill in the blank with the appropriate phrase.

> The relation f^{-1} is a function if and only if every horizontal line hits the graph of f ________________.

Explain why this statement is true.

Theorem 8.76. *If $f : X \to Y$ is a bijection, then*

(a) $f^{-1} \circ f = i_X$, *and*

(b) $f \circ f^{-1} = i_Y$.

Theorem 8.77. *If $f : X \to Y$ is a bijection, then $f^{-1} : Y \to X$ is also a bijection.*

Theorem 8.78. *If $f : X \to Y$ and $g : Y \to X$ are functions such that $g \circ f = i_X$ and $f \circ g = i_Y$, then f^{-1} is a function and $g = f^{-1}$.*

The upshot of Theorems 8.76 and 8.78 is that if f^{-1} is a function, then it is the only one satisfying the two-sided inverse property exhibited in Corollary 8.69 and Theorem 8.76. That is, inverse functions are unique when they exist. When the relation f^{-1} is a function, we call it the **inverse function** of f.

Problem 8.79. Let $X \subseteq \mathbb{R}$ and suppose $f : X \to \mathbb{R}$ is a function. Explain the difference between $f^{-1}(x)$ and $[f(x)]^{-1}$. When does each exist?

Problem 8.80. Let $X, Y \subseteq \mathbb{R}$ and define $f : X \to Y$ via $f(x) = e^x$ and $g : Y \to X$ via $g(x) = \ln(x)$. Identify the largest possible choices for X and Y so that f and g are inverses of each other.

Theorem 8.81. *If $f : X \to Y$ is a bijection, then $(f^{-1})^{-1} = f$.*

In the previous theorem, we restricted our attention to bijections so that f^{-1} would be a function, thus making $(f^{-1})^{-1}$ a sensible inverse relation in light of Definition 8.70. If we had defined inverses for arbitrary relations, then we would not have needed to require the function in Theorem 8.81 to be a bijection. In fact, we do not even need to require the relation to be a function. That is, if R is a relation from X to Y, then $(R^{-1})^{-1} = R$, as expected. Similarly, the next result generalizes to arbitrary relations.

Theorem 8.82. *If $f : X \to Y$ and $g : Y \to Z$ are both bijections, then $(g \circ f)^{-1} = f^{-1} \circ g^{-1}$.*

The previous theorem is sometimes referred to as the "socks and shoes theorem". Do you see how it got this name?

> The most difficult thing is the decision to act. The rest is merely tenacity.
>
> Amelia Earhart, aviation pioneer

8.4 Images and Preimages of Functions

There are two important types of sets related to functions.

Definition 8.83. Let $f : X \to Y$ be a function.

(a) If $S \subseteq X$, the **image** of S under f is defined via

$$\boxed{f(S) := \{f(x) \mid x \in S\}}.$$

(b) If $T \subseteq Y$, the **preimage** (or **inverse image**) of T under f is defined via

$$\boxed{f^{-1}(T) := \{x \in X \mid f(x) \in T\}}.$$

The image of a subset S of the domain is simply the subset of the codomain we obtain by mapping the elements of S. It is important to emphasize that the function f maps *elements* of X to *elements* of Y, but we can apply f to a *subset* of X to yield a *subset* of Y. That is, if $S \subseteq X$, then $f(S) \subseteq Y$. Note that the image of the domain is the same as the range of the function. That is, $f(X) = \text{Rng}(f)$.

When it comes to preimages, there is a real opportunity for confusion. In Section 8.3, we introduced the inverse relation f^{-1} of a function f (see Defintion 8.70) and proved that this relation is a function exactly when f is a bijection (see Theorem 8.74). If $f^{-1} : Y \to X$ is a function, then it is sensible to write $f^{-1}(y)$ for $y \in Y$. Notice that we defined the preimage of a subset of the codomain regardless of whether f^{-1} is a function or not. In particular, for $T \subseteq Y$, $f^{-1}(T)$ is the set of elements in the domain that map to elements in T. As a special case, $f^{-1}(\{y\})$ is the set of elements in the domain that map to $y \in Y$. If $y \notin \text{Rng}(f)$, then $f^{-1}(\{y\}) = \emptyset$. Notice that if $y \in Y$, $f^{-1}(\{y\})$ is always a sensible thing to write while $f^{-1}(y)$ only makes sense if f^{-1} is a function. Also, note that the preimage of the codomain is the domain. That is, $f^{-1}(Y) = X$.

Problem 8.84. Define $f : \mathbb{Z} \to \mathbb{Z}$ via $f(x) = x^2$. List elements in each of the following sets.

(a) $f(\{0, 1, 2\})$

(b) $f^{-1}(\{0, 1, 4\})$

Problem 8.85. Define $f : \mathbb{R} \to \mathbb{R}$ via $f(x) = 3x^2 - 4$. Find each of the following sets.

(a) $f(\{-1, 1\})$

(b) $f([-2, 4])$

(c) $f((-2, 4))$

(d) $f^{-1}([-10, 1])$

(e) $f^{-1}((-3, 3))$

(f) $f(\emptyset)$

(g) $f(\mathbb{R})$

(h) $f^{-1}(\{-1\})$

(i) $f^{-1}(\emptyset)$

(j) $f^{-1}(\mathbb{R})$

Problem 8.86. Define $f : \mathbb{R} \to \mathbb{R}$ via $f(x) = x^2$.

(a) Find two nonempty subsets A and B of $\mathbb{R}$ such that $A \cap B = \emptyset$ but $f^{-1}(A) = f^{-1}(B)$.

(b) Find two nonempty subsets A and B of $\mathbb{R}$ such that $A \cap B = \emptyset$ but $f(A) = f(B)$.

Problem 8.87. Suppose $f : X \to Y$ is an injection and A and B are disjoint subsets of X. Are $f(A)$ and $f(B)$ necessarily disjoint subsets of Y? If so, prove it. Otherwise, provide a counterexample.

Problem 8.88. Find examples of functions f and g together with sets S and T such that $f(f^{-1}(T)) \neq T$ and $g^{-1}(g(S)) \neq S$.

Problem 8.89. Let $f : X \to Y$ be a function and suppose $A, B \subseteq X$ and $C, D \subseteq Y$. Determine whether each of the following statements is true or false. If a statement is true, prove it. Otherwise, provide a counterexample.

(a) If $A \subseteq B$, then $f(A) \subseteq f(B)$.

(b) If $C \subseteq D$, then $f^{-1}(C) \subseteq f^{-1}(D)$.

(c) $f(A \cup B) \subseteq f(A) \cup f(B)$.

(d) $f(A \cup B) \supseteq f(A) \cup f(B)$.

(e) $f(A \cap B) \subseteq f(A) \cap f(B)$.

(f) $f(A \cap B) \supseteq f(A) \cap f(B)$.

(g) $f^{-1}(C \cup D) \subseteq f^{-1}(C) \cup f^{-1}(D)$.

(h) $f^{-1}(C \cup D) \supseteq f^{-1}(C) \cup f^{-1}(D)$.

(i) $f^{-1}(C \cap D) \subseteq f^{-1}(C) \cap f^{-1}(D)$.

(j) $f^{-1}(C \cap D) \supseteq f^{-1}(C) \cap f^{-1}(D)$.

(k) $A \subseteq f^{-1}(f(A))$.

(l) $A \supseteq f^{-1}(f(A))$.

(m) $f(f^{-1}(C)) \subseteq C$.

(n) $f(f^{-1}(C)) \supseteq C$.

Problem 8.90. For each of the statements in the previous problem that were false, determine conditions, if any, on the corresponding sets that would make the statement true.

We can generalize the results above to handle arbitrary collections of sets.

Theorem 8.91. *Let $f : X \to Y$ be a function and suppose $\{A_\alpha\}_{\alpha\in\Delta}$ is a collection of subsets of X.*

(a) $f\left(\bigcup_{\alpha\in\Delta} A_\alpha\right) = \bigcup_{\alpha\in\Delta} f(A_\alpha)$.

(b) $f\left(\bigcap_{\alpha\in\Delta} A_\alpha\right) \subseteq \bigcap_{\alpha\in\Delta} f(A_\alpha)$.

Theorem 8.92. *Let $f : X \to Y$ be a function and suppose $\{C_\alpha\}_{\alpha\in\Delta}$ is a collection of subsets of Y.*

(a) $f^{-1}\left(\bigcup_{\alpha\in\Delta} C_\alpha\right) = \bigcup_{\alpha\in\Delta} f^{-1}(C_\alpha)$.

(b) $f^{-1}\left(\bigcap_{\alpha\in\Delta} C_\alpha\right) = \bigcap_{\alpha\in\Delta} f^{-1}(C_\alpha)$.

Problem 8.93. Consider the equivalence relation given in Theorem 8.44. Explain why each equivalence class $[a]$ is equal to $f^{-1}(\{f(a)\})$.

Problem 8.94. Suppose that $f : \mathbb{R} \to \mathbb{R}$ is a function satisfying $f(x + y) = f(x) + f(y)$ for all $x, y \in \mathbb{R}$.

(a) Prove that $f(0) = 0$.

(b) Prove that $f(-x) = -f(x)$ for all $x \in \mathbb{R}$.

(c) Prove that f is injective if and only if $f^{-1}(\{0\}) = \{0\}$.

(d) Certainly every function given by $f(x) = mx$ for $m \in \mathbb{R}$ satisfies the initial hypothesis. Can you provide an example of a function that satisfies $f(x+y) = f(x) + f(y)$ that is not of the form $f(x) = mx$?

> The obstacle is the path.
>
> Zen saying, Author Unknown

8.5 Continuous Real Functions

In this section, we will explore the concept of continuity, which you likely encountered in high school.

Definition 8.95. A **real function** is any function $f : A \to \mathbb{R}$ such that A is a nonempty subset of $\mathbb{R}$.

There are several equivalent definitions of continuity for real functions. The following characterization is typically referred to as the **epsilon-delta definition of continuity**. Our definition mimics the definition of continuity used in metric spaces, which $\mathbb{R}$ equipped with absolute value happens to be an example of. Recall that $|a - b| < r$ means that the distance between a and b is less than r (see discussion below Corollary 5.31).

Definition 8.96. Suppose f is a real function such that $a \in \text{Dom}(f)$. We say that f is **continuous at** a if for every $\varepsilon > 0$, there exists $\delta > 0$ such that if $x \in \text{Dom}(f)$ and $|x - a| < \delta$, then $|f(x) - f(a)| < \varepsilon$. If f is continuous at every point in $B \subseteq \text{Dom}(f)$, then we say that f is **continuous on** B. If f is continuous on its entire domain, we simply say that f is **continuous**.

Loosely speaking, a real function f is continuous at the point $a \in \text{Dom}(f)$ if we can get $f(x)$ arbitrarily close to $f(a)$ by considering all $x \in \text{Dom}(f)$ sufficiently close to a. The value ε is indicating how close to $f(a)$ we need to be while the value δ is providing the "window" around a needed to guarantee that all points in the window (and in the domain) yield outputs within ε of $f(a)$. Figure 8.5 illustrates our definition of continuity. Note that in the figure, the point a is fixed while we need to consider all $x \in \text{Dom}(f)$ such that $|x - a| < \delta$. The dashed box in the figure has dimensions 2δ by 2ε and is centered at the point $(a, f(a))$. Intuitively, the function is continuous at a since given $\varepsilon > 0$, we could find $\delta > 0$ so that the graph of the function never exits the top or bottom of the dashed box.

Perhaps you have encountered the phrase "a function is continuous if you can draw its graph without lifting your pencil." While this description provides some intuition about what continuity of a function means, it is neither accurate nor precise enough to capture the meaning of continuity.

When proving that a function is continuous at a point, the choice of δ depends on both the point in question and the value of ε. An example should be helpful.

Example 8.97. Define $f : \mathbb{R} \to \mathbb{R}$ via $f(x) = 3x + 2$. Let's prove that f is continuous (at every point in the domain). Let $a \in \mathbb{R}$ and let $\varepsilon > 0$. Choose $\delta = \varepsilon/3$. We will see in a moment why this is a good choice for δ. Suppose $x \in \mathbb{R}$ such that $|x - a| < \delta$. We see that

$$|f(x) - f(a)| = |(3x + 2) - (3a + 2)| = |3x - 3a| = 3 \cdot |x - a| < 3 \cdot \delta = 3 \cdot \varepsilon/3 = \varepsilon.$$

We have shown that f is continuous at a, and since a was arbitrary, f is continuous.

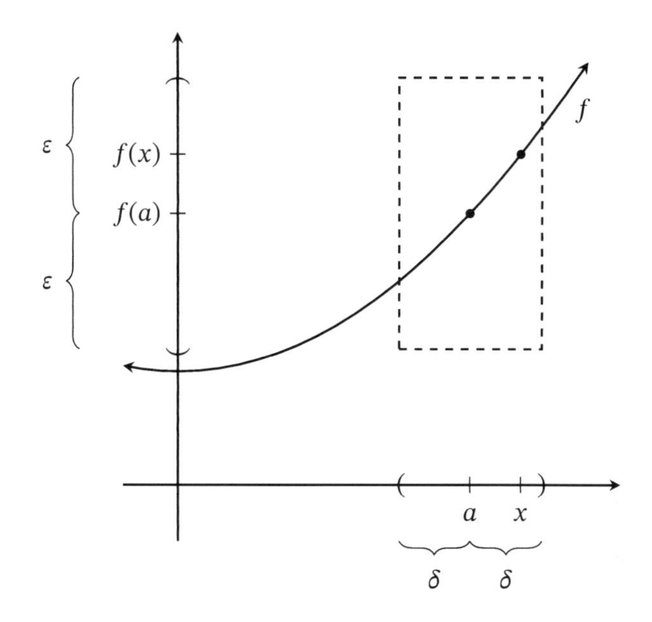

Figure 8.5. Visual representation of continuity of f at a.

Problem 8.98. Prove that each of the following real functions is continuous using Definition 8.96.

(a) $f : \mathbb{R} \to \mathbb{R}$ defined via $f(x) = x$.

(b) $g : \mathbb{R} \to \mathbb{R}$ defined via $g(x) = x + 42$.

(c) $h : \mathbb{R} \to \mathbb{R}$ defined via $h(x) = 5x$.

The next result tells us that every linear real function is continuous. Do not forget to handle the case when $m = 0$ in your proof. Note that the case when $m = 0$ proves that every constant function is continuous.

Theorem 8.99. *If $f : \mathbb{R} \to \mathbb{R}$ is defined via $f(x) = mx + b$ for $m, b \in \mathbb{R}$, then f is continuous.*

The second part of the next problem is much harder than you might expect.

Problem 8.100. Define $f : \mathbb{R} \to \mathbb{R}$ via $f(x) = x^2$.

(a) Prove that f is continuous at 0.

(b) Prove that f is continuous at 1.

Problem 8.101. Define $f : \mathbb{R} \to \mathbb{R}$ via $f(x) = \sqrt{x}$. Prove that f is continuous at 0.

Problem 8.102. Suppose f is a real function. Write a precise statement for what it means for f to not be continuous at $a \in \text{Dom}(f)$.

Problem 8.103. Define $f : \mathbb{R} \to \mathbb{R}$ via

$$f(x) = \begin{cases} 1, & \text{if } x = 0 \\ x, & \text{otherwise.} \end{cases}$$

Determine where f is continuous and justify your assertion.

Problem 8.104. Define $f : \mathbb{R} \to \mathbb{R}$ via

$$f(x) = \begin{cases} 1, & \text{if } x \in \mathbb{Q} \\ 0, & \text{otherwise.} \end{cases}$$

Determine where f is continuous and justify your assertion.

After completing the next problem, reflect on the statement "a function is continuous if you can draw its graph without lifting your pencil."

Problem 8.105. Define $f : \mathbb{N} \to \mathbb{R}$ via $f(x) = 1$. Notice the domain! Determine where f is continuous and justify your assertion.

Theorem 8.106. *Suppose f is a real function. Then f is continuous if and only if the preimage $f^{-1}(U)$ of every open set U is an open set intersected with the domain of f.*

The previous characterization of continuity is often referred to as the "open set definition of continuity," although for us it is a theorem instead of a definition. This is the definition used in topology. Another notion of continuity, called "sequential continuity", makes use of convergent sequences. All of these characterizations of continuity are equivalent for the real numbers (using the standard definition of an open set). However, there are contexts in mathematics where the epsilon-delta definition of continuity is undefined (because there is not a notion of distance in either the domain or codomain) and others where continuity and sequential continuity are not equivalent.

Since every open set is the union of bounded open intervals (Definition 5.53), the union of open sets is open (Theorem 5.58), and preimages respect unions (Theorem 8.92), we can strengthen Theorem 8.106 into a slightly more useful result.

Theorem 8.107. *Suppose f is a real function. Then f is continuous if and only if the preimage $f^{-1}(I)$ of every bounded open interval I is an open set intersected with the domain of f.*

Now that we have two methods for verifying continuity (Definition 8.96 and Theorem 8.106/8.107), you can use either one when approaching the remaining problems in this section. Sometimes it does not matter which approach you take and other times one method might be better suited to the task.

Problem 8.108. Define $f : \mathbb{R} \to \mathbb{R}$ via $f(x) = x^2$. Prove that f is continuous.

Problem 8.109. Define $f : \mathbb{R} \setminus \{0\} \to \mathbb{R}$ via $f(x) = \frac{1}{x}$. Determine where f is continuous and justify your assertion.

The previous problems once again calls into question the phrase "a function is continuous if you can draw its graph without lifting your pencil."

Problem 8.110. Find a continuous real function f and an open interval I such that the preimage $f^{-1}(I)$ is not an open interval.

For the next few problems, if you attempt to construct counterexamples, you may rely on your previous knowledge about various functions that you encountered in high school and calculus.

Problem 8.111. Suppose f is a continuous real function. If U is an open set contained in $\text{Dom}(f)$, is the image $f(U)$ always open? If so, prove it. Otherwise, provide a counterexample.

Problem 8.112. Suppose f is a continuous real function. If C is a closed set, is the preimage $f^{-1}(C)$ always a closed set? If so, prove it. Otherwise, provide a counterexample.

Problem 8.113. Suppose f is a continuous real function. If $[a, b]$ is a closed interval contained in $\text{Dom}(f)$, is the image $f([a, b])$ always a closed interval? If so, prove it. Otherwise, provide a counterexample.

Problem 8.114. Suppose f is a continuous real function. If C is a closed set contained in $\text{Dom}(f)$, is the image $f(C)$ always a closed set? If so, prove it. Otherwise, provide a counterexample.

Problem 8.115. Suppose f is a continuous real function. If B is bounded set contained in $\text{Dom}(f)$, is the image $f(B)$ always a bounded set? If so, prove it. Otherwise, provide a counterexample.

Problem 8.116. Suppose f is a continuous real function. If B is a bounded set, is the preimage $f^{-1}(B)$ always a bounded set? If so, prove it. Otherwise, provide a counterexample.

Problem 8.117. Suppose f is a continuous real function. If K is a compact set, is the preimage $f^{-1}(B)$ always a compact set? If so, prove it. Otherwise, provide a counterexample.

Problem 8.118. Suppose f is a continuous real function. If C is a connected set contained in $\text{Dom}(f)$, is the image $f(C)$ always connected? If so, prove it. Otherwise, provide a counterexample.

Problem 8.119. Suppose f is a continuous real function. If C is a connected set, is the preimage $f^{-1}(C)$ always a connected set? If so, prove it. Otherwise, provide a counterexample.

Perhaps you noticed the absence of one natural question in the previous sequence of problems. If f is a continuous real function and K is a subset of the domain of f, is the image $f(K)$ a compact set? It turns out that the answer is "yes", but proving this fact is beyond the scope of this book. This theorem is often proved in a real analysis course and is then used to prove the Extreme Value Theorem, which you may have encountered in your calculus course.

The next result is a special case of the well-known **Intermediate Value Theorem**, which states that if f is a continuous real function whose domain contains the interval $[a, b]$, then f attains every value between $f(a)$ and $f(b)$ at some point within the interval $[a, b]$. To prove the special case, utilize Theorem 5.87 and Problem 8.118 together with a proof by contradiction.

Theorem 8.120. *Suppose f is a real function. If f is continuous on $[a, b]$ such that $f(a) < 0 < f(b)$ or $f(a) > 0 > f(b)$, then there exists $r \in [a, b]$ such that $f(r) = 0$.*

If we generalize the previous result, we obtain the Intermediate Value Theorem.

Theorem 8.121 (Intermediate Value Theorem)**.** *Suppose f is a real function. If f is continuous on $[a, b]$ such that $f(a) < c < f(b)$ or $f(a) > c > f(b)$ for some $c \in \mathbb{R}$, then there exists $r \in [a, b]$ such that $f(r) = c$.*

Problem 8.122. Is the converse of the Intermediate Value Theorem true? If so, prove it. Otherwise, provide a counterexample.

> The miracle of the appropriateness of the language of mathematics for the formulation of the laws of physics is a wonderful gift which we neither understand nor deserve. We should be grateful for it and hope that it will remain valid in future research and that it will extend, for better or for worse, to our pleasure, even though perhaps also to our bafflement, to wide branches of learning.
>
> Eugene Paul Wigner, theoretical physicist

I counted everything. I counted the steps to the road, the steps up to church, the number of dishes and silverware I washed ... anything that could be counted, I did.

Katherine Johnson, mathematician

9

Cardinality

In this chapter, we will explore the notion of cardinality, which formalizes what it means for two sets to be the same "size".

9.1 Introduction to Cardinality

What does it mean for two sets to have the same "size"? If the sets are finite, this is easy: just count how many elements are in each set. Another approach would be to pair up the elements in each set and see if there are any left over. In other words, check to see if there is a one-to-one correspondence (i.e., bijection) between the two sets.

But what if the sets are infinite? For example, consider the set of natural numbers $\mathbb{N}$ and the set of even natural numbers $2\mathbb{N} := \{2n \mid n \in \mathbb{N}\}$. Clearly, $2\mathbb{N}$ is a proper subset of $\mathbb{N}$. Moreover, both sets are infinite. In this case, you might be thinking that $\mathbb{N}$ is "larger than" $2\mathbb{N}$. However, it turns out that there is a one-to-one correspondence between these two sets. In particular, consider the function $f : \mathbb{N} \to 2\mathbb{N}$ defined via $f(n) = 2n$. It is easily verified that f is both injective and surjective. In this case, mathematics has determined that the right viewpoint is that $\mathbb{N}$ and $2\mathbb{N}$ do have the same "size". However, it is clear that "size" is a bit too imprecise when it comes to infinite sets. We need something more rigorous.

Definition 9.1. Let A and B be sets. We say that A and B have the same **cardinality** if there exists a bijection between A and B. In this case, we write $\boxed{\operatorname{card}(A) = \operatorname{card}(B)}$.

Note that we have not defined $\operatorname{card}(A)$ by itself. Doing so would not be too difficult for finite sets, but making such a notation precise in general is tricky business. When we write $\text{card}(A) = \text{card}(B)$ (and later $\text{card}(A) \leq \text{card}(B)$ and $\text{card}(A) < \text{card}(B)$), we are asserting the existence of a certain type of function from A to B.

If f is a bijection from A to B, then by Theorem 8.77, f^{-1} is a bijection from B to A. Either one of these functions can be utilized to prove that $\text{card}(A) = \text{card}(B)$. This idea is worth keeping in mind as you tackle problems in this chapter. In particular, you might have an easier time creating a bijection between two sets in one direction over the other. This is often a limitation of the human mind as to opposed to some fundamental mathematical difficulty.

Example 9.2. Let $A = \{1, 2, 3, 4, 5\}$ and $B = \{\text{apple, banana, cherry, dragon fruit, elderberry}\}$. The function $f : A \to B$ given by

$$f = \{(1, \text{apple}), (2, \text{banana}), (3, \text{cherry}), (4, \text{dragon fruit}), (5, \text{elderberry})\}$$

is a bijection from A to B, and hence $\text{card}(A) = \text{card}(B)$. Note that this is not the only bijection from A to B. In fact, there are $5! = 120$ bijections from A to B, one of which is the function f we defined above. The inverse of each bijection from A to B is a bijection from B to A. We could also use any of of these bijections to verify that $\text{card}(A) = \text{card}(B)$.

Example 9.3. Define $f : \mathbb{Z} \to 6\mathbb{Z}$ via $f(n) = 6n$. It is easily verified that f is both injective and surjective, and hence $\text{card}(\mathbb{Z}) = \text{card}(6\mathbb{Z})$. We could also utilize the inverse function $f^{-1} : 6\mathbb{Z} \to \mathbb{Z}$ given by $f^{-1}(n) = \frac{1}{6}n$ to show that $\mathbb{Z}$ and $6\mathbb{Z}$ have the same cardinality.

Example 9.4. Let $\mathbb{R}^+$ denote the set of positive real numbers and define $f : \mathbb{R} \to \mathbb{R}^+$ via $f(x) = e^x$. As you are likely familiar with, this exponential function is a bijection, and so $\text{card}(\mathbb{R}) = \text{card}(\mathbb{R}^+)$. Similar to the previous example, we could also use the inverse function $f^{-1} : \mathbb{R}^+ \to \mathbb{R}$ given by $f^{-1}(x) = \ln(x)$ to show that these two sets have the same cardinality.

The previous two examples illustrate an important distinction between finite sets and infinite sets, namely infinite sets can be in bijection with proper subsets of themselves! Theorems 9.23 and 9.31 will make this idea explicit.

Example 9.5. Let $m, n \in \mathbb{N} \cup \{0\}$. A North-East lattice path from $(0, 0)$ to (m, n) is path in the plane from $(0, 0)$ to (m, n) consisting only steps either one unit North or one unit East. Note that every lattice path from $(0, 0)$ to (m, n) consists of a total of $m+n$ steps. Figure 9.1 shows a North-East lattice path from $(0, 0)$ to $(4, 3)$. Let $\mathcal{L}_{m,n}$ denote the set of North-East paths from $(0, 0)$ to (m, n). For example, the North-East lattice path given in Figure 9.1 is an element of $\mathcal{L}_{4,3}$. A binary string of length k is an ordered list of consisting of k entries where each entry is either 0 or 1. For example, 0101100 and 0101001 are two different binary strings of length 7. Let $\mathcal{S}_k$ denote the set of binary strings of length k. For example, $\mathcal{S}_3 = \{000, 100, 010, 001, 110, 101, 011, 111\}$. We claim that there is a bijection between $\mathcal{L}_{m,n}$ and $\mathcal{S}_{m+n}$. One such bijection is given by mapping a lattice path to the string that results by assigning each East step to 0 and each North step to 1 as we travel the path from $(0,0)$ to (m,n). Using this construction, the lattice path in Figure 9.1 would get mapped to the binary string 0101100. Since no two lattice paths will map to the same string, our mapping is injective. Given a string in $\mathcal{S}_{m+n}$, it is easy to find the lattice path in $\mathcal{L}_{m,n}$ that maps to it, and so our function is also surjective. Thus, our mapping is a bijection between $\mathcal{L}_{m,n}$ and $\mathcal{S}_{m+n}$. We have shown that $\text{card}(\mathcal{L}_{m,n}) = \text{card}(\mathcal{S}_{m+n})$.

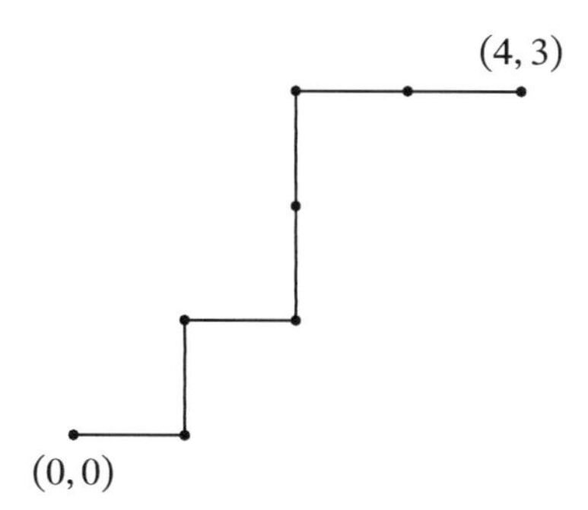

Figure 9.1. A North-East lattice path from $(0,0)$ to $(4,3)$.

When approaching Part (d) of the next problem, try creating a linear function $f : (a,b) \to (c,d)$. Drawing a picture should help.

Problem 9.6. Prove each of the following. In each case, you should create a bijection between the two sets. Briefly justify that your functions are in fact bijections.

(a) $\text{card}(\{a,b,c\}) = \text{card}(\{x,y,z\})$

(b) $\text{card}(\mathbb{N}) = \text{card}(\{2n+1 \mid n \in \mathbb{N}\})$

(c) $\text{card}(\mathbb{N}) = \text{card}(\mathbb{Z})$

(d) $\text{card}((a,b)) = \text{card}((c,d))$ (where (a,b) and (c,d) are intervals)

(e) $\text{card}(\mathbb{N}) = \text{card}\left(\left\{\frac{1}{2^n} \mid n \in \mathbb{N}\right\}\right)$

Problem 9.7. If A is a set, do A and $A \times \{x\}$ have the same cardinality? Justify your answer.

Problem 9.8. Let $\mathcal{D}_n$ denote the collection of North-East lattice paths from $(0,0)$ to (n,n) that never drop below the line $y = x$. These types of lattice paths are often called Dyck paths after the German mathematician Walther Franz Anton von Dyck (1856–1934). A sequence of parentheses is balanced if it can be parsed syntactically. In other words, there should be the same number of open parentheses "(" and closed parentheses ")", and when reading from left to right there should never be more closed parentheses than open. For example, ()()() and ()(()) are balanced parenthesizations consisting of three pairs of parentheses while ())(() and ()(()(are not balanced. Let $\mathcal{B}_n$ denote the collection of balanced parenthesizations consisting of n pairs of parentheses. For example, $\mathcal{B}_3 = \{()()(), ()(()), (()()), (())(), ((()))\}$.

(a) Find all Dyck paths in $\mathcal{D}_3$.

(b) Prove that $\operatorname{card}(\mathcal{D}_n) = \operatorname{card}(\mathcal{B}_n)$.

For Part (b) of the next problem, start by defining $\varphi : \mathcal{F} \to \mathcal{P}(\mathbb{N})$ so that $\varphi(f)$ yields a subset of $\mathbb{N}$ determined by when f outputs a 1.

Problem 9.9. Let $\mathcal{F}$ denote the set of functions from $\mathbb{N}$ to $\{0,1\}$.

(a) Describe at least three functions in $\mathcal{F}$.

(b) Prove that $\mathcal{F}$ and $\mathcal{P}(\mathbb{N})$ have the same cardinality.

Our first theorem concerning cardinality will likely not come as a surprise.

Theorem 9.10. *Let A, B, and C be sets.*

(a) $\operatorname{card}(A) = \operatorname{card}(A)$.

(b) *If* $\operatorname{card}(A) = \operatorname{card}(B)$, *then* $\operatorname{card}(B) = \operatorname{card}(A)$.

(c) *If* $\operatorname{card}(A) = \operatorname{card}(B)$ *and* $\operatorname{card}(B) = \operatorname{card}(C)$, *then* $\operatorname{card}(A) = \operatorname{card}(C)$.

In light of the previous theorem, the next result should not be surprising.

Corollary 9.11. *If X is a set, then "has the same cardinality as" is an equivalence relation on $\mathcal{P}(X)$.*

Theorem 9.12. *Let A, B, C, and D be sets such that* $\operatorname{card}(A) = \operatorname{card}(C)$ *and* $\operatorname{card}(B) = \operatorname{card}(D)$.

(a) *If A and B are disjoint and C and D are disjoint, then* $\operatorname{card}(A \cup B) = \operatorname{card}(C \cup D)$.

(b) $\operatorname{card}(A \times B) = \operatorname{card}(C \times D)$.

Given two finite sets, it makes sense to say that one set is "larger than" another provided one set contains more elements than the other. We would like to generalize this idea to handle both finite and infinite sets.

Definition 9.13. Let A and B be sets. If there is an injective function from A to B, then we say that the **cardinality of A is less than or equal to the cardinality of B**. In this case, we write $\boxed{\operatorname{card}(A) \leq \operatorname{card}(B)}$.

Theorem 9.14. *Let A, B, and C be sets.*

(a) *If $A \subseteq B$, then* $\operatorname{card}(A) \leq \operatorname{card}(B)$.

(b) *If* $\operatorname{card}(A) \leq \operatorname{card}(B)$ *and* $\operatorname{card}(B) \leq \operatorname{card}(C)$, *then* $\operatorname{card}(A) \leq \operatorname{card}(C)$.

(c) *If $C \subseteq A$ while* $\operatorname{card}(B) = \operatorname{card}(C)$, *then* $\operatorname{card}(B) \leq \operatorname{card}(A)$.

It might be tempting to think that the existence of injective function from a set A to a set B that is *not* surjective would verify that $\operatorname{card}(A) \leq \operatorname{card}(B)$ and $\operatorname{card}(A) \neq \operatorname{card}(B)$. While this is true for finite sets, it is not true for infinite sets as the next problem asks you to verify.

Problem 9.15. Provide an example of sets A and B such that $\operatorname{card}(A) = \operatorname{card}(B)$ despite the fact that there exists an injective function from A to B that is not surjective.

Definition 9.16. Let A and B be sets. We write $\boxed{\operatorname{card}(A) < \operatorname{card}(B)}$ if $\operatorname{card}(A) \leq \operatorname{card}(B)$ and $\operatorname{card}(A) \neq \operatorname{card}(B)$.

As a reminder, the statements $\operatorname{card}(A) = \operatorname{card}(B)$ and $\operatorname{card}(A) \leq \operatorname{card}(B)$ are symbolic ways of asserting the existence of certain types of functions from A to B. When we write $\operatorname{card}(A) < \operatorname{card}(B)$, we are saying something much stronger than "There exists a function $f : A \to B$ that is injective but not surjective." The statement $\operatorname{card}(A) < \operatorname{card}(B)$ is asserting that *every* injective function from A to B is not surjective. In general, it is difficult to prove statements like $\operatorname{card}(A) \neq \operatorname{card}(B)$ or $\operatorname{card}(A) < \operatorname{card}(B)$.

> You will become clever through your mistakes.
>
> German Proverb

9.2 Finite Sets

In the previous section, we used the phrase "finite set" without formally defining it. Let's be a bit more precise. The following shorthand comes in handy.

Definition 9.17. For each $n \in \mathbb{N}$, define $[n] := \{1, 2, \dots, n\}$.

For example, $[5] = \{1, 2, 3, 4, 5\}$. Notice that our notation looks just like the notation we used for equivalence classes. However, despite the similar notation, these concepts are unrelated. We will have to rely on context to keep them straight.

The next definition should coincide with your intuition about what it means for a set to be finite.

Definition 9.18. A set A is **finite** if $A = \emptyset$ or $\operatorname{card}(A) = \operatorname{card}([n])$ for some $n \in \mathbb{N}$. If $A = \emptyset$, then we say that A has **cardinality** 0 and if $\operatorname{card}(A) = \operatorname{card}([n])$, then we say that A has **cardinality** n.

Let's prove a few results about finite sets. When proving the following theorems, do not forget to consider the empty set.

Theorem 9.19. *If A is finite and $\operatorname{card}(A) = \operatorname{card}(B)$, then B is finite.*

Theorem 9.20. *If A has cardinality $n \in \mathbb{N} \cup \{0\}$ and $x \notin A$, then $A \cup \{x\}$ is finite and has cardinality $n + 1$.*

Consider using induction when proving the next theorem.

Theorem 9.21. *For every $n \in \mathbb{N}$, every subset of $[n]$ is finite.*

Theorem 9.20 shows that adding a single element to a finite set increases the cardinality by 1. As you would expect, removing one element from a finite set decreases the cardinality by 1.

Theorem 9.22. *If A has cardinality $n \in \mathbb{N}$, then for all $x \in A$, $A \setminus \{x\}$ is finite and has cardinality $n - 1$.*

The next result tells us that the cardinality of a proper subset of a finite set is never the same as the cardinality of the original set. It turns out that this theorem does not hold for infinite sets.

Theorem 9.23. *Every subset of a finite set is finite. In particular, if A is a finite set, then $\operatorname{card}(B) < \operatorname{card}(A)$ for all proper subsets B of A.*

Induction is a sensible approach to proving the next two theorems.

Theorem 9.24. *If $A_1, A_2, \dots, A_k$ is a finite collection of finite sets, then $\displaystyle\bigcup_{i=1}^{k} A_i$ is finite.*

The next theorem, called the **Pigeonhole Principle**, is surprisingly useful. It puts restrictions on when we may have an injective function. The name of the theorem is inspired by the following idea: If n pigeons wish to roost in a house with k pigeonholes and $n > k$, then it must be the case that at least one hole contains more than one pigeon. Note that 2 is the smallest value of n that makes sense in the hypothesis below.

Theorem 9.25 (Pigeonhole Principle). *If $n, k \in \mathbb{N}$ and $f : [n] \to [k]$ with $n > k$, then f is not injective.*

> God created infinity, and man, unable to understand infinity, had to invent finite sets.
>
> ---
>
> Gian-Carlo Rota, mathematician & philosopher

9.3 Infinite Sets

In the previous section, we explored finite sets. Now, let's tinker with infinite sets.

Definition 9.26. A set A is **infinite** if A is not finite.

Let's see if we can utilize this definition to prove that the set of natural numbers is infinite. For sake of a contradiction, assume otherwise. Then there exists $n \in \mathbb{N}$ such that $\mathrm{card}([n]) = \mathrm{card}(\mathbb{N})$, which implies that there exists a bijection $f : [n] \to \mathbb{N}$. What can you say about the number

$$m := \max(f(1), f(2), \dots, f(n)) + 1?$$

Theorem 9.27. *The set $\mathbb{N}$ of natural numbers is infinite.*

The next theorem is analogous to Theorem 9.19, but for infinite sets. To prove this theorem, try a proof by contradiction. You should end up composing two bijections, say $f : A \to B$ and $g : B \to [n]$ for some $n \in \mathbb{N}$. As we shall see later, the converse of this theorem is not true in general.

Theorem 9.28. *If A is infinite and $\mathrm{card}(A) = \mathrm{card}(B)$, then B is infinite.*

Problem 9.29. Quickly verify that the following sets are infinite by appealing to Theorem 9.27, Theorem 9.28, or Problem 9.6.

(a) The set of odd natural numbers

(b) The set of even natural numbers

(c) $\mathbb{Z}$

(d) $R = \left\{\frac{1}{2^n} \mid n \in \mathbb{N}\right\}$

(e) $\mathbb{N} \times \{a\}$

Notice that Definition 9.26 tells us what infinite sets are not, but it doesn't really tell us what they are. In light of Theorem 9.27, one way of thinking about infinite sets is as follows. Suppose A is some nonempty set. Let's select a random element from A and set it aside. We will call this element the "first" element. Then we select one of the remaining elements and set is aside, as well. This is the "second" element. Imagine we continue this way, choosing a "third" element, and "fourth" element, etc. If the set is infinite, we should never run out of elements to select. Otherwise, we would create a bijection with $[n]$ for some $n \in \mathbb{N}$.

The next problem, sometimes referred to as the Hilbert Hotel, named after the German mathematician David Hilbert (1862–1942), illustrates another way of thinking about infinite sets.

Problem 9.30. The Infinite Hotel has rooms numbered $1, 2, 3, 4, \ldots$. Every room in the Infinite Hotel is currently occupied.

(i) Is it possible to make room for one more guest (assuming they want a room all to themselves)?

(ii) An infinite number of new guests, say $g_1, g_2, g_3, \ldots$, show up in the lobby and each demands a room. Is it possible to make room for all the new guests even if the hotel is already full?

The previous problem verifies that there exists a proper subset of the natural numbers that is in bijection with the natural numbers themselves. We also witnessed this in Parts (a) and (b) of Problem 9.29. Notice that Theorem 9.23 forbids this type of behavior for finite sets. It turns out that this phenomenon is true for all infinite sets. The next theorem verifies that that the two viewpoints of infinite sets discussed above are valid. To prove this theorem, we need to prove that the three statements are equivalent. One possible approach is to prove (i) if and only if (ii) and (ii) if and only if (iii). For (i) implies (ii), construct f recursively. For (ii) implies (i), try a proof by contradiction. For (ii) implies (iii), let $B = A \setminus \{f(1), f(2), \ldots\}$ and show that A can be put in bijection with $B \cup \{f(2), f(3), \ldots\}$. Lastly, for (iii) implies (ii), suppose $g : A \to C$ is a bijection for some proper subset C of A. Let $a \in A \setminus C$. Define $f : \mathbb{N} \to A$ via $f(n) = g^n(a)$, where g^n means compose g with itself n times.

Theorem 9.31. *The following statements are equivalent.*

(i) *The set A is infinite.*

(ii) *There exists an injective function $f : \mathbb{N} \to A$.*

(iii) *The set A can be put in bijection with a proper subset of A (i.e., there exists a proper subset B of A such that* $\text{card}(B) = \text{card}(A)$*).*

It is worth mentioning that for the previous theorem, (iii) implies (i) follows immediately from the contrapositive of Theorem 9.23. When proving (i) implies (ii) in the previous theorem, did you apply the Axiom of Choice? If so, where?

Corollary 9.32. *A set is infinite if and only if it has an infinite subset.*

Corollary 9.33. *If A is an infinite set, then* $\text{card}(\mathbb{N}) \leq \text{card}(A)$.

Problem 9.34. Find a new proof of Theorem 9.27 that uses (iii) implies (i) from Theorem 9.31.

Problem 9.35. Quickly verify that the following sets are infinite by appealing to either Theorem 9.31 (use (ii) implies (i)) or Corollary 9.32.

(a) Set of odd natural numbers

(b) Set of even natural numbers

(c) $\mathbb{Z}$

(d) $\mathbb{N} \times \mathbb{N}$

(e) $\mathbb{Q}$

(f) $\mathbb{R}$

(g) Set of perfect squares in $\mathbb{N}$

(h) $(0, 1)$

(i) $\mathbb{C} := \{a + bi \mid a, b \in \mathbb{R}\}$

> An ounce of practice is worth more than tons of preaching.
>
> ---
>
> Mahatma Gandhi, political activist

9.4 Countable Sets

Recall that if $A = \emptyset$, then we say that A has cardinality 0. Also, if $\text{card}(A) = \text{card}([n])$ for $n \in \mathbb{N}$, then we say that A has cardinality n. We have a special way of describing sets that are in bijection with the natural numbers.

Definition 9.36. If A is a set such that $\text{card}(A) = \text{card}(\mathbb{N})$, then we say that A is **denumerable** and has **cardinality** $\aleph_0$ (read "aleph naught").

Notice if a set A has cardinality $1, 2, \ldots,$ or $\aleph_0$, we can label the elements in A as "first", "second", and so on. That is, we can "count" the elements in these situations. Certainly, if a set has cardinality 0, counting is not an issue. This idea leads to the following definition.

Definition 9.37. A set A is called **countable** if A is finite or denumerable. A set is called **uncountable** if it is not countable.

Problem 9.38. Quickly justify that each of the following sets is countable. Feel free to appeal to previous problems. Which sets are denumerable?

(a) $\{a, b, c\}$

(b) Set of odd natural numbers

(c) Set of even natural numbers

(d) $\left\{\frac{1}{2^n} \mid n \in \mathbb{N}\right\}$

(e) Set of perfect squares in $\mathbb{N}$

(f) $\mathbb{Z}$

(g) $\mathbb{N} \times \{a\}$

Utilize Theorem 9.31 or Corollary 9.33 when proving the next result.

Theorem 9.39. *Every infinite set contains a denumerable subset.*

Theorem 9.40. *Let A and B be sets such that A is countable. If $f : A \to B$ is a bijection, then B is countable.*

For the next proof, consider the cases when A is finite versus infinite. The contrapositive of Corollary 9.32 should be useful for the case when A is finite.

Theorem 9.41. *Every subset of a countable set is countable.*

Theorem 9.42. *A set is countable if and only if it has the same cardinality of some subset of the natural numbers.*

Theorem 9.43. *If $f : \mathbb{N} \to A$ is a surjective function, then A is countable.*

Loosely speaking, the next theorem tells us that we can arrange all of the rational numbers then count them "first", "second", "third", etc. Given the fact that between any two distinct rational numbers on the number line, there are an infinite number of other rational numbers (justified by taking repeated midpoints), this may seem counterintuitive.

Here is one possible approach for proving the next theorem. Make a table with column headings $0, 1, -1, 2, -2, \ldots$ and row headings $1, 2, 3, 4, 5, \ldots$. If a column has heading m and a row has heading n, then the entry in the table corresponds to the fraction m/n. Find a way to zig-zag through the table making sure to hit every entry in the table (not including column and row headings) exactly once. This justifies that there is a bijection between $\mathbb{N}$ and the entries in the table. Do you see why? But now notice that every rational number appears an infinite number of times in the table. Resolve this by appealing to Theorem 9.41.

Theorem 9.44. *The set of rational numbers $\mathbb{Q}$ is countable.*

Theorem 9.45. *If A and B are countable sets, then $A \cup B$ is countable.*

We would like to prove a stronger result than the previous theorem. To do so, we need an intermediate result.

Theorem 9.46. *Let $\{A_n\}_{n=1}^{\infty}$ be a collection of sets. Define $B_1 := A_1$ and for each natural number $n > 1$, define*

$$B_n := A_n \setminus \bigcup_{i=1}^{n-1} A_i.$$

Then we we have the following:

(a) *The collection $\{B_n\}_{n=1}^{\infty}$ is pairwise disjoint.*

(b) $\displaystyle\bigcup_{n=1}^{\infty} A_n = \bigcup_{n=1}^{\infty} B_n.$

The next theorem states that the union of a countable collection of countable sets is countable. To prove this, consider two cases:

(1) The collection of sets is finite.

(2) The collection of sets is infinite.

To handle the first case, use induction together with Theorem 9.45. The second case is substantially more challenging. First, use Theorem 9.46 to construct a collection $\{B_n\}$ of pairwise disjoint sets whose union is equal to the union of the original collection. Since each B_n is a subset of one of the sets in the original collection and each of these sets is countable, each B_n is also countable by Theorem 9.41. This implies that for each n, we can write $B_n = \{b_{n,1}, b_{n,2}, b_{n,3}, \ldots\}$, where the set may be finite or infinite. From here, we outline two different approaches for continuing. One approach is to construct a bijection from $\mathbb{N}$ to $\bigcup_{n=1}^{\infty} B_n$ using Figure 9.2 as inspiration. One thing you will need to address is what to do when a set in the collection $\{B_n\}$ is finite. For the second approach, define $f : \bigcup_{n=1}^{\infty} B_n \to \mathbb{N}$ via $f(b_{n,m}) = 2^n 3^m$, verify that this function is injective, and then appeal to Theorem 9.41. Try using both of these approaches when tackling the proof of the following theorem.

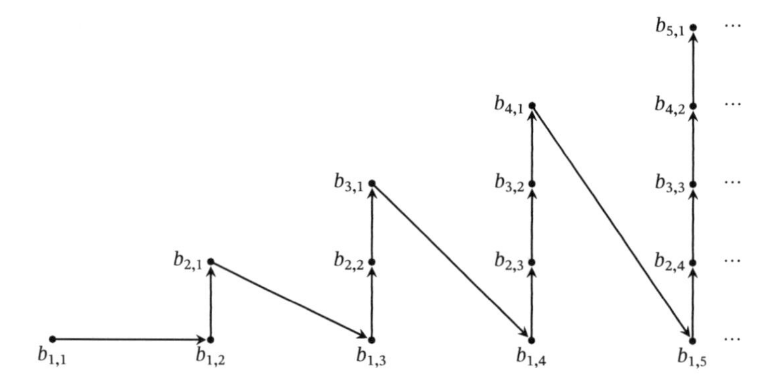

Figure 9.2. Inspiration for one possible approach to proving Theorem 9.47.

Theorem 9.47. *Let Δ be equal to either $\mathbb{N}$ or $[k]$ for some $k \in \mathbb{N}$. If $\{A_n\}_{n\in\Delta}$ is a countable collection of sets such that each A_n is countable, then $\bigcup_{n\in\Delta} A_n$ is countable.*

Did you use the Axiom of Choice when proving the previous theorem? If so, where?

Theorem 9.48. *If A and B are countable sets, then $A \times B$ is countable.*

Theorem 9.49. *The set of all finite sequences of 0's and 1's (e.g., 0110010 is a finite sequence consisting of 0's and 1') is countable.*

Theorem 9.50. *The collection of all finite subsets of a countable set is countable.*

Vulnerability is not winning or losing; it's having the courage to show up and be seen when we have no control over the outcome.

Brené Brown, storyteller & author

9.5 Uncountable Sets

Recall from Definition 9.37 that a set A is uncountable if A is not countable. Since all finite sets are countable, the only way a set could be uncountable is if it is infinite. It follows that a set A is uncountable if and only if there is never a bijection between $\mathbb{N}$ and A. It is not clear that uncountable sets even exist! It turns out that uncountable sets do exist and in this section, we will discover a few of them.

Our first task is to prove that the interval $(0, 1)$ is uncountable. By Problem 9.35(h), we know that $(0, 1)$ is an infinite set, so it is at least plausible that $(0, 1)$ is uncountable. The following problem outlines the proof of Theorem 9.52. Our approach is often referred to as **Cantor's Diagonalization Argument**, named after German mathematician Georg Cantor (1845–1918).

Before we get started, recall that every number in $(0, 1)$ can be written in decimal form. However, there may be more than one way to write a given number in decimal form. For example, 0.2 equals $0.1\overline{99}$. A number $0.a_1a_2a_3\ldots$ in $(0, 1)$ is said to be in **standard decimal form** if there is no k such that for all $i > k$, $a_i = 9$. That is, a number is in standard decimal form if and only if its decimal expansion does not end with a repeating sequence of 9's. For example, 0.2 is in standard decimal form while $0.1\overline{99}$ is not, even though both represent the same number. It turns out that every real number can be expressed uniquely in standard decimal form. We will take this fact for granted.

Problem 9.51. For sake of a contradiction, assume the interval $(0, 1)$ is countable. Then there exists a bijection $f : \mathbb{N} \to (0, 1)$. For each $n \in \mathbb{N}$, its image under f is some number in $(0, 1)$. Write $f(n) = 0.a_{1n}a_{2n}a_{3n}\ldots$, where a_{1n} is the first digit in the standard decimal form for the image of n, a_{2n} is the second digit, and so on. If $f(n)$ terminates after k digits, then our convention will be to continue the decimal expansion with 0's. Now, define $b = 0.b_1b_2b_3\ldots$, where

$$b_i = \begin{cases} 2, & \text{if } a_{ii} \neq 2 \\ 3, & \text{if } a_{ii} = 2. \end{cases}$$

(a) Prove that the decimal expansion that defines b above is in standard decimal form.

(b) Prove that for all $n \in \mathbb{N}$, $f(n) \neq b$.

(c) Explain why f cannot be surjective and why this is a contradiction.

You just proved that the interval $(0, 1)$ cannot be countable!

The previous problem proves following theorem.

Theorem 9.52. *The open interval* $(0, 1)$ *is uncountable.*

Loosely speaking, what Theorem 9.52 says is that the open interval $(0, 1)$ is "bigger" in terms of the number of elements it contains than the natural numbers and even the rational numbers. This shows that there are infinite sets of different sizes! We now know there is at least one uncountable set, namely the interval $(0, 1)$. The next three results are useful for finding other uncountable sets. For the first theorem, try a proof by contradiction and take a look at Theorem 9.41.

Theorem 9.53. *If A and B are sets such that $A \subseteq B$ and A is uncountable, then B is uncountable.*

Corollary 9.54. *If A and B are sets such that A is uncountable and B is countable, then $A \setminus B$ is uncountable.*

Theorem 9.55. *If $f : A \to B$ is an injective function and A is uncountable, then B is uncountable.*

Since the interval $(0, 1)$ is uncountable and $(0, 1) \subseteq \mathbb{R}$, it follows immediately from Theorem 9.53 that $\mathbb{R}$ is also uncountable. The next theorem tells that $(0, 1)$ and $\mathbb{R}$ actually have the same cardinality! To prove this, consider the function $f : (0, 1) \to \mathbb{R}$ defined via $f(x) = \tan(\pi x - \frac{\pi}{2})$.

Theorem 9.56. *The real numbers are uncountable. In particular,* $\text{card}((0, 1)) = \text{card}(\mathbb{R})$.

The **continuum hypothesis**—originally proposed by Cantor in 1878—states that there is no set whose cardinality is strictly between that of the natural numbers and the real numbers. Cantor unsuccessfully attempted to prove the continuum hypothesis for several years. It follows from the work of Paul Cohen (1934–2007) and Kurt Gödel (1906–1978) that the continuum hypothesis and its negation are independent of the Zermelo-Fraenkel axioms of set theory (briefly discussed at the end of Section 3.2). That is, either the continuum hypothesis or its negation can be added as an axiom to ZFC set theory, with the resulting theory being consistent if and only if ZFC is consistent (i.e., no contradictions are produced). Nowadays, most set theorists believe that the continuum hypothesis *should* be false.

Theorem 9.57. *If $a, b \in \mathbb{R}$ with $a < b$, then (a, b), $[a, b]$, $(a, b]$, and $[a, b)$ are all uncountable.*

Theorem 9.58. *The set of irrational numbers is uncountable.*

Theorem 9.59. *The set $\mathbb{C}$ of complex numbers is uncountable.*

Problem 9.60. Determine whether each of the following statements is true or false. If a statement is true, prove it. Otherwise, provide a counterexample.

(a) If A and B are sets such that A is uncountable, then $A \cup B$ is uncountable.

(b) If A and B are sets such that A is uncountable, then $A \cap B$ is uncountable.

(c) If A and B are sets such that A is uncountable, then $A \times B$ is uncountable.

(d) If A and B are sets such that A is uncountable, then $A \setminus B$ is uncountable.

An approach similar to Cantor's Diagonalization Argument will be helpful when approaching the next problem.

Problem 9.61. Let S be the set of infinite sequences of 0's and 1's. Determine whether S is countable or uncountable and prove that your answer is correct.

Theorem 9.62. *If S is the set from Problem* 9.61, *then* $\text{card}(\mathcal{P}(\mathbb{N})) = \text{card}(S)$.

Corollary 9.63. *The power set of the natural numbers is uncountable.*

Notice that $\mathbb{N}$ is countable while $\mathcal{P}(\mathbb{N})$ is uncountable. That is, the power set of the natural numbers has cardinality strictly larger than the natural numbers. We generalize this phenomenon in the next theorem.

According to Theorem 9.56 and Corollary 9.63, $\mathbb{R}$ and $\mathcal{P}(\mathbb{N})$ are both uncountable. In fact, $\text{card}(\mathcal{P}(\mathbb{N})) = \text{card}(\mathbb{R})$, which we state without proof. However, it turns out that the two uncountable sets may or may not have the same cardinality. Perhaps surprisingly, there are sets that are even "bigger" than the set of real numbers. The next theorem is named after Georg Cantor, who first stated and proved it at the end of the 19th century. The theorem states that given any set, we can always increase the cardinality by considering its power set. That is, if A is a set, $\mathcal{P}(A)$ has strictly greater cardinality than A itself. For finite sets, Cantor's theorem follows from Theorems 4.11 and 4.12 (both of which we proved via induction). Perhaps much more surprising is that Cantor discovered an elegant argument that is applicable to any set, whether finite or infinite. To prove Cantor's Theorem, first exhibit an injective function from A to $\mathcal{P}(A)$. This proves that $\text{card}(A) \leq \text{card}(\mathcal{P}(A))$. To show that $\text{card}(A) < \text{card}(\mathcal{P}(A))$, try a proof by contradiction. That is, assume there exists a bijective function $f : A \to \mathcal{P}(A))$. Derive a contradiction by considering the set $B = \{x \in A \mid x \notin f(x)\}$.

Theorem 9.64 (Cantor's Theorem). *If A is a set, then* $\operatorname{card}(A) < \operatorname{card}(\mathcal{P}(A))$.

Recall that cardinality provides a way for talking about "how big" a set is. The fact that the natural numbers and the real numbers have different cardinality (one countable, the other uncountable), tells us that there are at least two different "sizes of infinity". By iteratively taking the power set of an infinite set and applying Cantor's Theorem we obtain an endless hierarchy of cardinalities, each strictly larger than the one before it. Colloquially, this implies that there are "infinitely many sizes of infinity" and there is "no largest infinity".

> If you want to sharpen a sword, you have to remove a little metal.
>
> ---
>
> Author Unknown

Appendix A

Elements of Style for Proofs

Mathematics is about discovering proofs and writing them clearly and compellingly. The following guidelines apply whenever you write a proof. Keep these guidelines handy so that you may refer to them as you write your proofs.

(1) **The burden of communication lies on you, not on your reader.** It is your job to explain your thoughts; it is not your reader's job to guess them from a few hints. You are trying to convince a skeptical reader who does not believe you, so you need to argue with airtight logic in crystal clear language; otherwise the reader will continue to doubt. If you did not write something on the paper, then (a) you did not communicate it,(b) the reader did not learn it, and (c) the grader has to assume you did not know it in the first place.

(2) **Tell the reader what you are proving or citing.** The reader does not necessarily know or remember what "Theorem 2.13" is. Even a professor grading a stack of papers might lose track from time to time. Therefore, the statement you are proving should be on the same page as the beginning of your proof.

In most proofs you will want to refer to an earlier definition, problem, theorem, or corollary. In this case, you should reference the statement by number, but it is also helpful to the reader to summarize the statement you are citing. For example, you might write something like, "By Theorem 2.3, the sum of two consecutive integers is odd, and so...."

(3) **Use English words.** Although there will usually be equations or mathematical statements in your proofs, use English sentences to connect them

and display their logical relationships. If you at proofs in textbooks and research papers, you will see that they consist mostly of English words.

(4) **Use complete sentences.** If you wrote a history essay in sentence fragments, the reader would not understand what you meant; likewise in mathematics you must use complete sentences, with verbs, to convey your logical train of thought.

Some complete sentences can be written purely in mathematical symbols, such as equations (e.g., $a^3 = b^{-1}$), inequalities (e.g., $x < 5$), and other relations (like $5|10$ or $7 \in \mathbb{Z}$). These statements usually express a relationship between two mathematical *objects*, like numbers or sets. However, it is considered bad style to begin a sentence with symbols. A common phrase to use to avoid starting a sentence with mathematical symbols is "We see that...".

(5) **Show the logical connections among your sentences.** Use phrases like "Therefore", "Thus", "Hence", "Then", "since", "because", "if..., then...", or "if and only if" to connect your sentences.

(6) **Know the difference between statements and objects.** A mathematical object is a *thing*, a noun, such as a set, an element, a number, an ordered pair, a vector space, etc. Objects either exist or do not exist. Statements, on the other hand, are mathematical *sentences*: they are either true or false.

When you see or write a cluster of math symbols, be sure you know whether it is an object (e.g., "$x^2 + 3$") or a statement (e.g., "$x^2 + 3 < 7$"). One way to tell is that every mathematical statement includes a verb, such as $=$, $\leq$, $\in$, "divides", etc.

(7) **The symbol "=" means "equals".** Do not write $A = B$ unless you mean that A actually equals B. This guideline seems obvious, but there is a great temptation to be sloppy. In calculus, for example, some people might write $f(x) = x^2 = 2x$ (which is false), when they really mean that "if $f(x) = x^2$, then $f'(x) = 2x$."

(8) **Do not interchange = and $\Longrightarrow$.** The equals sign connects two *objects*, as in "$x^2 = b$"; the symbol "$\Longrightarrow$" is an abbreviation for "implies" and connects two *statements*, as in "$a + b = a \Longrightarrow b = 0$." You should avoid using $\Longrightarrow$ in formal write-ups of proofs.

(9) **Avoid logical symbols in your proofs.** Similar to $\Longrightarrow$, you should avoid using the logical symbols $\forall$, $\exists$, $\vee$, $\wedge$, and $\Longleftrightarrow$ in your formal write-ups. These symbols are useful for abbreviating in your scratch work.

(10) **Say exactly what you mean.** Just as = is sometimes abused, so too people sometimes write $A \in B$ when they mean $A \subseteq B$, or write $a_{ij} \in A$ when they mean that a_{ij} is an entry in matrix A. Mathematics is a very precise language, and there is a way to say exactly what you mean; find it and use it.

(11) **Do not utilize anything unproven.** Every statement in your proof should be something you *know* to be true. The reader expects your proof to be a series of statements, each proven by the statements that came before it. If you ever need to write something you do not yet know is true, you *must* preface it with words like "assume," "suppose," or "if" if you are temporarily assuming it, or with words like "we need to show that" or "we claim that" if it is your goal. Otherwise, the reader will think they have missed part of your proof.

(12) **Write strings of equalities (or inequalities) in the proper order.** When your reader sees something like

$$A = B \leq C = D,$$

they expect to understand easily why $A = B$, why $B \leq C$, and why $C = D$, and they expect the point of the entire line to be the more complicated fact that $A \leq D$. For example, if you were computing the distance d of the point $(12, 5)$ from the origin, you could write

$$d = \sqrt{12^2 + 5^2} = 13.$$

In this string of equalities, the first equals sign is true by the Pythagorean theorem, the second is just arithmetic, and the conclusion is that the first item equals the last item: $d = 13$.

A common error is to write strings of equations in the wrong order. For example, if you were to write "$\sqrt{12^2 + 5^2} = 13 = d$", your reader would understand the first equals sign, would be baffled as to how we know $d = 13$, and would be utterly perplexed as to why you wanted or needed to go through 13 to prove that $\sqrt{12^2 + 5^2} = d$.

(13) **Avoid circularity.** Be sure that no step in your proof makes use of the conclusion!

(14) **Do not write the proof backwards.** Beginning students often attempt to write "proofs" like the following, which attempts to prove that

$\tan^2(x) = \sec^2(x) - 1$:

$$
\begin{aligned}
\tan^2(x) &= \sec^2(x) - 1 \\
\left(\frac{\sin(x)}{\cos(x)}\right)^2 &= \frac{1}{\cos^2(x)} - 1 \\
\frac{\sin^2(x)}{\cos^2(x)} &= \frac{1 - \cos^2(x)}{\cos^2(x)} \\
\sin^2(x) &= 1 - \cos^2(x) \\
\sin^2(x) + \cos^2(x) &= 1 \\
1 &= 1
\end{aligned}
$$

Notice what has happened here: the student *started* with the conclusion, and deduced the true statement "$1 = 1$." In other words, they have proved "If $\tan^2(x) = \sec^2(x) - 1$, then $1 = 1$," which is true but highly uninteresting.

Now this is not a bad way of *finding* a proof. Working backwards from your goal often is a good strategy *on your scratch paper*, but when it is time to *write* your proof, you have to start with the hypotheses and work to the conclusion.

Here is an example of a suitable proof for the desired result, where each expression follows from the one immediately proceeding it:

$$
\begin{aligned}
\sec^2(x) - 1 &= \frac{1}{\cos^2(x)} - 1 \\
&= \frac{1 - \cos^2(x)}{\cos^2(x)} \\
&= \frac{\sin^2(x)}{\cos^2(x)} \\
&= \left(\frac{\sin(x)}{\cos(x)}\right)^2 \\
&= (\tan(x))^2 \\
&= \tan^2(x).
\end{aligned}
$$

(15) **Be concise.** Many beginning proof writers err by writing their proofs too short, so that the reader cannot understand their logic. It is nevertheless quite possible to be too wordy, and if you find yourself writing a full-page essay, it is possible that you do not really have a proof, but just some intuition. When you find a way to turn that intuition into a formal proof, it will be much shorter.

(16) **Introduce every symbol you use.** If you use the letter "k," the reader should know exactly what k is. Good phrases for introducing symbols include "Let $n \in \mathbb{N}$," "Let k be the least integer such that...," "For every real number a...," and "Suppose $A \subseteq \mathbb{R}$...".

(17) **Use appropriate quantifiers (once).** When you introduce a variable $x \in S$, it must be clear to your reader whether you mean "for all $x \in S$" or just "for some $x \in S$." If you just say something like "$y = x^2$ where $x \in S$," the word "where" does not indicate whether you mean "for all" or "some".

Phrases indicating the quantifier "for all" include "Let $x \in S$"; "for all $x \in S$"; "for every $x \in S$"; "for each $x \in S$"; etc. Phrases indicating the quantifier "some" or "there exists") include "for some $x \in S$"; "there exists an $x \in S$"; "for a suitable choice of $x \in S$"; etc.

Once you have said "Let $x \in S$," the letter x has its meaning defined. You do not need to say "for all $x \in S$" again, and you definitely should *not* say "let $x \in S$" again.

(18) **Use a symbol to mean only one thing.** Once you use the letter x once, its meaning is fixed for the duration of your proof. You cannot use x to mean anything else. There is an exception to this guideline. Sometimes a proof will include multiple subproofs that are distinct from each other. In this case, you can reuse a variable or symbol as long as it is clear to the reader that you have concluded with the previous subproof and have moved onto a new subproof.

(19) **Do not "prove by example."** Most problems ask you to prove that something is true "for all"—You *cannot* prove this by giving a single example, or even a hundred. Your proof will need to be a logical argument that holds for *every example there possibly could be.*

On the other hand, if the claim that you are trying to prove involves the existence of a mathematical object with a particular property, then providing a specific example is sufficient.

(20) **Write "Let $x = \ldots$," not "Let $\cdots = x$."** When you have an existing expression, say a^2, and you want to give it a new, simpler name like b, you should write "Let $b = a^2$," which means, "Let the new symbol b mean a^2." This convention makes it clear to the reader that b is the brand-new symbol and a^2 is the old expression he/she already understands.

If you were to write it backwards, saying "Let $a^2 = b$," then your startled reader would ask, "What if $a^2 \neq b$?"

(21) **Make your counterexamples concrete and specific.** Proofs need to be entirely general, but counterexamples should be concrete. When you provide an example or counterexample, make it as specific as possible. For a

set, for example, you must specify its elements, and for a function you must specify the corresponding relation (possibly an algebraic rule) and its domain and codomain. Do not say things like "f could be one-to-one but not onto"; instead, provide an actual function f that is one-to-one but not onto.

(22) **Do not include examples in proofs.** Including an example very rarely adds anything to your proof. If your logic is sound, then it does not need an example to back it up. If your logic is bad, a dozen examples will not help it (see Guideline 19). There are only two valid reasons to include an example in a proof: if it is a *counterexample* disproving something, or if you are performing complicated manipulations in a general setting and the example is just to help the reader understand what you are saying.

(23) **Use scratch paper.** Finding your proof will be a long, potentially messy process, full of false starts and dead ends. Do all that on scratch paper until you find a real proof, and only then break out your clean paper to write your final proof carefully.

Only sentences that actually contribute to your proof should be part of the proof. Do not just perform a "brain dump," throwing everything you know onto the paper before showing the logical steps that prove the conclusion. *That is what scratch paper is for.*

Appendix B

Fancy Mathematical Terms

Here are some important mathematical terms that you will encounter throughout mathematics.

(1) **Definition**—a precise and unambiguous description of the meaning of a mathematical term. It characterizes the meaning of a word by giving all the properties and only those properties that must be true.

(2) **Theorem**—a mathematical statement that is proved using rigorous mathematical reasoning. In a mathematical paper, the term theorem is often reserved for the most important results.

(3) **Proposition**—a proved and often interesting result, but generally less important than a theorem.

(4) **Lemma**—a minor result whose sole purpose is to help in proving a theorem. It is a stepping stone on the path to proving a theorem. Occasionally lemmas can take on a life of their own (Zorn's Lemma, Urysohn's Lemma, Burnside's Lemma, Sperner's Lemma).

(5) **Corollary**—a result in which the (usually short) proof relies heavily on a given theorem (we often say that "this is a corollary of Theorem A").

(6) **Conjecture**—a statement that is unproved, but is believed to be true (Collatz Conjecture, Goldbach Conjecture, Twin prime Conjecture).

(7) **Claim**—an assertion that is then proved. It is often used like an informal lemma.

(8) **Counterexample**—a specific example showing that a statement is false.

(9) **Axiom/Postulate**—a statement that is assumed to be true without proof. These are the basic building blocks from which all theorems are proved (Euclid's five postulates, axioms of ZFC, Peano axioms).

(10) **Identity**—a mathematical expression giving the equality of two (often variable) quantities (trigonometric identities, Euler's identity).

(11) **Paradox**—a statement that can be shown, using a given set of axioms and definitions, to be both true and false. Paradoxes are often used to show the inconsistencies in a flawed axiomatic theory (e.g., Russell's Paradox). The term paradox is also used informally to describe a surprising or counterintuitive result that follows from a given set of rules (Banach-Tarski Paradox, Alabama Paradox, Gabriel's Horn).

Appendix C

Paradoxes

A **paradox** is a statement that can be shown, using a given set of axioms and definitions, to be both true and false. Recall that an axiom is a statement that is assumed to be true without proof. These are the basic building blocks from which all theorems are proved. Paradoxes are often used to show the inconsistencies in a flawed axiomatic theory. The term paradox is also used informally to describe a surprising or counterintuitive result that follows from a given set of rules. In Section 3.2, we encountered two paradoxes:

- The Barber of Seville (Problem 3.24)
- Russell's Paradox (Problem 3.26)

Below are several additional paradoxes that are worth exploring.

(1) **Librarian's Paradox.** A librarian is given the unenviable task of creating two new books for the library. Book A contains the names of all books in the library that reference themselves and Book B contains the names of all books in the library that do not reference themselves. But the librarian just created two new books for the library, so their titles must be in either Book A or Book B. Clearly Book A can be listed in Book B, but where should the librarian list Book B?

(2) **Liar's Paradox.** Consider the statement: this sentence is false. Is it true or false?

(3) **Berry Paradox.** Consider the claim: every natural number can be unambiguously described in fourteen words or less. It seems clear that this statement is false, but if that is so, then there is some smallest natural number

which cannot be unambiguously described in fourteen words or less. Let's call it n. But now n is "the smallest natural number that cannot be unambiguously described in fourteen words or less." This is a complete and unambiguous description of n in fourteen words, contradicting the fact that n was supposed not to have such a description. Therefore, all natural numbers can be unambiguously described in fourteen words or less!

(4) **The Naming Numbers Paradox.** Consider the claim: every natural number can be unambiguously described using no more than 50 characters (where a character is a–z, 0–9, and a "space"). For example, we can describe 9 as "9" or "nine" or "the square of the second prime number." There are only 37 characters, so we can describe at most 37^{50} numbers, which is very large, but not infinite. So the statement is false. However, here is a "proof" that it is true. Let S be the set of natural numbers that can be unambiguously described using no more than 50 characters. For the sake of contradiction, suppose it is not all of $\mathbb{N}$. Then there is a smallest number $t \in \mathbb{N} \setminus S$. We can describe t as: the smallest natural number not in S. Thus t can be described using no more than 50 characters. So $t \in S$, a contradiction.

(5) **Euathlus and Protagoras.** Euathlus wanted to become a lawyer but could not pay Protagoras. Protagoras agreed to teach him under the condition that if Euathlus won his first case, he would pay Protagoras, otherwise not. Euathlus finished his course of study and did nothing. Protagoras sued for his fee. He argued:

If Euathlus loses this case, then he must pay (by the judgment of the court).
If Euathlus wins this case, then he must pay (by the terms of the contract).
He must either win or lose this case.
Therefore Euathlus must pay me.

But Euathlus had learned well the art of rhetoric. He responded:

If I win this case, I do not have to pay (by the judgment of the court).
If I lose this case, I do not have to pay (by the contract).
I must either win or lose the case.
Therefore, I do not have to pay Protagoras.

Appendix D

Definitions in Mathematics

It is difficult to overstate the importance of definitions in mathematics. Definitions play a different role in mathematics than they do in everyday life.

Suppose you give your friend a piece of paper containing the definition of the rarely-used word **rodomontade**. According to the Oxford English Dictionary[1] (OED) it is:

> A vainglorious brag or boast; an extravagantly boastful, arrogant, or bombastic speech or piece of writing; an arrogant act.

Give your friend some time to study the definition. Then take away the paper. Ten minutes later ask her to define rodomontade. Most likely she will be able to give a reasonably accurate definition. Maybe she'd say something like, "It is a speech or act or piece of writing created by a pompous or egotistical person who wants to show off how great they are." It is unlikely that she will have quoted the OED word-for-word. In everyday English that is fine—you would probably agree that your friend knows the meaning of the rodomontade. This is because most definitions are *descriptive*. They describe the common usage of a word.

Let us take a mathematical example. The OED[2] gives this definition of **continuous**.

> Characterized by continuity; extending in space without interruption of substance; having no interstices or breaks; having its parts in immediate connection; connected, unbroken.

Likewise, we often hear calculus students speak of a continuous function as one whose graph can be drawn "without picking up the pencil." This definition is

[1] http://www.oed.com/view/Entry/166837
[2] http://www.oed.com/view/Entry/40280

descriptive. However, as we learned in calculus, the picking-up-the-pencil description is not a perfect description of continuous functions. This is not a mathematical definition.

Mathematical definitions are *prescriptive*. The definition must prescribe the exact and correct meaning of a word. Contrast the OED's descriptive definition of continuous with the definition of continuous found in a real analysis textbook.

> A function $f : A \to \mathbb{R}$ is **continuous at a point** $c \in A$ if, for all $\varepsilon > 0$, there exists $\delta > 0$ such that whenever $|x - c| < \delta$ (and $x \in A$) it follows that $|f(x) - f(c)| < \varepsilon$. If f is continuous at every point in the domain A, then we say that f is **continuous on** A.[3]

In mathematics there is very little freedom in definitions. Mathematics is a deductive theory; it is impossible to state and prove theorems without clear definitions of the mathematical terms. The definition of a term must completely, accurately, and unambiguously describe the term. Each word is chosen very carefully and the order of the words is critical. In the definition of continuity changing "there exists" to "for all," changing the orders of quantifiers, changing $<$ to $\leq$ or $>$, or changing $\mathbb{R}$ to $\mathbb{Z}$ would completely change the meaning of the definition.

What does this mean for you, the student? Our recommendation is that at this stage you memorize the definitions word-for-word. It is the safest way to guarantee that you have it correct. As you gain confidence and familiarity with the subject you may be ready to modify the wording. You may want to change "for all" to "given any" or you may want to change $|x - c| < \delta$ to $-\delta < x - c < \delta$ or to "the distance between x and c is less than δ."

Of course, memorization is not enough; you must have a conceptual understanding of the term, you must see how the formal definition matches up with your conceptual understanding, and you must know how to work with the definition. It is perhaps with the first of these that descriptive definitions are useful. They are useful for building intuition and for painting the "big picture." Only after days (weeks, months, years?) of experience does one get an intuitive feel for the epsilon-delta definition of continuity; most mathematicians have the "picking-up-the-pencil" definitions in their head. This is fine as long as we know that it is imperfect, and that when we prove theorems about continuous functions in mathematics we use the mathematical definition.

We end this discussion with an amusing real-life example in which a descriptive definition was not sufficient. In 2003 the German version of the game show *Who wants to be a millionaire?* contained the following question: "Every rectangle is: (a) a rhombus, (b) a trapezoid, (c) a square, (d) a parallelogram."

The confused contestant decided to skip the question and left with €4000. Afterward the show received letters from irate viewers. Why were the contestant

[3]This definition is taken from page 109 of Stephen Abbott's *Understanding Analysis*, but the definition would be essentially the same in any modern real analysis textbook.

and the viewers upset with this problem? Clearly a rectangle is a parallelogram, so (d) is the answer. But what about (b)? Is a rectangle a trapezoid? We would describe a trapezoid as a quadrilateral with a pair of parallel sides. But this leaves open the question: can a trapezoid have *two* pairs of parallel sides or must there only be *one* pair? The viewers said two pairs is allowed, the producers of the television show said it is not. This is a case in which a clear, precise, mathematical definition is required.